THE
GENOMIC
KITCHEN

THE GENOMIC KITCHEN

YOUR GUIDE TO UNDERSTANDING
AND USING THE FOOD-GENE CONNECTION
FOR A LIFETIME OF HEALTH

Amanda Archibald, RD

Book design by Blue Design
(www.bluedes.com)

Printed in the USA

About the Links and References

The Genomic Kitchen cites more than 100 books,
websites, and, particularly, peer-reviewed scientific
papers from all over the world. You can find the
citations at the end of the book, chapter by chapter
by topic in order of discussion. Particularly in the case
of scientific papers, I have cited the PubMed abstract.
Some citations will link directly to the primary source—
the actual article, book, or website; the majority will be
abstracts that allow you electronic access after paying
a small fee.

If you are using the print edition, you will note
words and phrases highlighted in red. These indicate
a citation at the end of the book which you can use via
search engine to reach primary source material.

Every graphic in this book relating to the
Genomic Kitchen Ingredient Toolbox—lists, charts,
and guides as well as many other resources-
-is downloadable from my website with the
URL: **https://www.genomickitchen.com/
genomic-kitchen-book-resources.**

Although I will update the book as needed,
sometimes links are broken when pages migrate,
sites are updated, or when publishers stop supporting
individual titles. When this happens, you can generally
use the word or phrase to find the source with your
preferred search engine.

"We all know you are what you eat...except it's more complicated than that. Every bite of food ... is a signal you're sending your genes to turn on or off, controlling your body composition, health, and longevity. If you care about your health, you need to read this amazing work. I can't wait to give this book to my patients."
— **Matt Dawson, M.D., Wild Health**

"Knowing better ways to get the nutrients your body needs to run better becomes deliciously easy with this book. The Genomic Kitchen is the resource we need to discover how our genes want us to eat for better health."
— **Ashley Koff RD, CEO, The Better Nutrition Program**

"Excellent work by a brilliant, forward-thinking dietitian."
— **Lisa Dorfman, MS, RD, CSSD, CCMS, LMHC, FAND**

"Amanda Archibald delivers a digestible primer for health through nutrition, based on cutting edge scientific research. This is the handbook we have been waiting for."
— **Mary Jo Fishburn M.D.**

"Amanda Archibald, RD has given the non-nutrigenomic scientist (which let's face it, is pretty much all of us) the Codex to not only decipher this missing piece in the food – health puzzle, but the tools to readily apply it to everyday cookery. Quite simply put, there is a reason this is a required text for my Culinary Medicine class!"
— **Michael S. Fenster, M.D., FACC, FSCA&I, PEMBA**

CONTENTS

Foreword

I have been a registered dietitian for more than twenty-five years. In that time, I have seen the pendulum swing from its being a highly measured and somewhat confining dictate on nutrition do's and don'ts to today's confusing mash-up of restrictive diets, outsized promises and conflicting information.

But both eras remain predicated on a one size fits all mentality, catering to a single objective or an isolated condition. Or even a single organ or body part! Having spent so many years in navigating the ever-changing turbulent waters of nutrition science, I instinctively knew there had to be a way to be kind to your body, helping it find its own best health, while continuing to enjoy food, cooking and a shared meal.

I was introduced to nutrigenomics while teaching at a conference in South Africa, and knew immediately that this was the guiding light I had been looking for. Finally a laser-focused tool that would separate nutrition myth from fact and provide precise answers to individuals about what to eat, how to prepare it, and, importantly, why.

Discovering nutrigenomics was an a-ha moment for me. I have

always believed in the power of nutritious ingredients made into delicious food to be shared in company to affect health and well-being. With the advent of genomic sequencing and the analysis of ingredients for factors that influence gene behavior, an entire world opened up for me – for all of us, really – that can revolutionize our definition of healthy eating. Not surprisingly, some of this science corroborates what we have already identified as longevity diets and healthy foods. What's unique to nutrigenomics – and the space I have defined as culinary genomics – is that we now have the knowledge to use key ingredients in ways that can not only assure their nutritional integrity but also in many instances amplify it.

This book does not require that you have personalized genomic testing. The principles I share are universal to specific, well-researched gene behaviors. I hope you will use this book as intended - as a tool for discovery. I invite you to use it as your guide to understanding how food impacts your genes, biochemistry and health as well as to using this knowledge to your inspired advantage to make and share food filled with life. These revolutionary new food truths are indeed a key to a lifetime of delicious, empowered health and wellbeing. I hope you will be as excited to acquaint yourself with culinary genomics as I have been to explore it.

Introduction

Genomic medicine offers a new approach to personalizing health care, one which uses your unique gene blueprint as powerful insight into how your genes respond to nutrients, exercise, life stressors, environmental pollutants and medications. It does not get any more personal than YOU. Nutrigenomics, which looks at nutrition through the lens of genes and how they are affected by what and how you eat, is an exciting, relatively young offshoot of the broader study of genetics. Both are highly complex fields that are rapidly evolving as researchers attempt to untangle the relationship between nutrition and the unique blueprint of our bodies, our genes. More specifically, the field explores how some classes of substances found in our diet can affect how genes express themselves at the molecular level to have a positive or negative effect on our health, our behavior, and even our emotions. For those of us who are involved in this revolution, these are indeed watershed times.

The promise of genomic medicine in the twenty-first century is that we will start to uncover the susceptibility of an individual to a disease or number of diseases based on that person's genomic blueprint. We all have a unique genomic blueprint.

Our individuality is expressed not only by our genes but also by the "spelling errors" or aberrations on them called Single Nucleotide Polymorphisms, or SNPs (pronounced "snips"). Using an individual's genomic information, we can apply lifestyle, nutrition, supplement, and exercise recommendations to influence gene behavior. We can complement this insight with the use of biomarkers from lab tests to measure the effectiveness of those recommendations and continue to adjust them. And although most of us have not had a personalized genomic test, that doesn't mean we can't apply evidence-based genomic science principles, both general and specific, to achieve better health. For those of you who want to dive deeper, Appendix A: "Genomic Testing. For me or Not For Me?" at the end of this book.

Genetics and Aging

One of the most exciting areas that genomic medicine elucidates is our understanding of which genes are associated with longevity—promoting a longer life. We all want to live a disease-free life without physical hindrance or pain.

Through research among centenarians around the world, researchers are uncovering genetic variations which may explain why some people live to be 80 years old while others live to be 110. "Being genetically so programmed," we learn in The Genetics of Exceptional Longevity, "primary aging is uncontrollable and irreversible whereas secondary aging is a biological process in which physical structure and biological function deteriorate over the years. This last process is susceptible to some control since it is mediated by lifestyle, social, and environmental factors."

Genomic research among centenarians identifies variants on specific genes which are conducive to longevity. A SNP—one of those spelling errors—on a gene called ACE, for example, may be one in a pattern of SNPs implicated in the onset of heart disease.

Another example is the ApoE gene which plays an influential role in promoting lipid (fat) transportation, protein metabolism, and injury repair in the brain. A SNP on this gene, coupled with other genomic factors, is associated with increased risk for developing cardiovascular disease and potentially Alzheimer's. Research among centenarians shows that a lower occurence of the ApoE e4-allele (a genetic variant) is often associated with a decreased incidence of cardiovascular disease and Alzheimer's. Likewise, genomic research allows us to identify patterns of gene variants that are positively associated with longevity, thus providing us with both signposts to look for and a path to follow: What might we be able to do to activate those genes and influence their actions in our own bodies?

These are just two examples of genes that researchers have identified in influencing the long life of centenarians. Such genomic research provides deep insights into how we may be able to manage disease onset related to genomic information and help live healthier longer lives. That's a goal we all want to achieve.

The Genomic Kitchen explores this dynamic world in a way that helps you make clear and deep connections between the food that you eat and its influence on the genes that play a role in the root cause of disease. This is not a cookbook, but you *will* discover new concepts for what to eat and how to cook. This is not a science text, but you *will* learn the basic concepts that explain the relationship between genes and nutrition.

Think of *The Genomic Kitchen* as a guide to a new way of thinking about your body and your health. I have focused on the leading health factors associated with aging and longevity—inflammation, oxidative stress (how your body handles troublesome free radicals), metabolism of fat and carbohydrates and the health of your gut and its microbiome—to provide you with a brand new culinary perspective that is backed by genomic science. The ingredients you will meet – or become better acquainted with – here can help anyone. While our individual genetic blueprint is unique, the principles you'll learn in The Genomic Kitchen apply to genes that all of us possess.

You could say it is just the beginning of your journey. We hope so.

Chapter One: Discovery in the Dolomites

Deep in the heart of the Italian Alps tucked into the quiet village of St. Vito di Cadore, two hours north of Venice, is AGA, a tiny restaurant participating in the "zero kilometer" food movement. This movement, also known as "KM 0," suggests a way of eating with zero additives and preservatives and 100-percent local ingredients, i.e., zero kilometers between production and plate.

I love this area the most in the summer—hiking the Sella Ronda, a loop trail around the Sella massif, backpacking from one *refugio* (mountain hut) to another, high up among the rocky pinnacles. Occasionally you'll round a corner and witness a mountain goat majestically perched on a jagged outcrop, peering across the deep valley to the mountains on the other side. On more daring occasions, I have climbed with my backpack, grasping at ropes on my way up precipitous rock faces. I confess that I've never had to courage to ascend the region's famous iron ladders up the steepest inclines, a section of trail known as the Via Ferrata. Perhaps next time.

I have lived or traveled widely in Europe, Africa, and the United

States. My friends call me a 'Restless Traveler,' because I'm always heading off on another adventure. I can't help it. While I experience a deep personal satisfaction in adventurous travel, there is also a professional element. As a nutritionist, I am fascinated by how geography and culture influence our relationship with food, and how food choices affect the way we live, feel, and act.

Many people might pass through San Vito di Cadore and drive right past AGA, a small restaurant situated on the ground floor of the Hotel Trieste, for this tiny village is not a destination but a waypoint to Cortina d'Ampezzo and the Dolomite resorts just beyond it. Get out of your car and the beauty of those towering mountains surrounds you. You can see man's work, too, the famous Cortina ski slopes, the Cinque Torri towers, even the old ski jump from the 1956 Winter Olympic Games.

There's not much in San Vito di Cadore but ski shops on the narrow streets and two *pasticceria* (pastry shops) thronged with *macchiato*-mad skiers on their way to or from the slopes in the winter. That and AGA, which has just four tables to seat, perhaps, twelve diners. It is a modest venue featuring two talented young chefs who serve a cuisine that may hold the secret to improving the way the world eats—and thereby its health.

A friend who happened to be living in Italy several years back first told me about AGA, this restaurant where I was to have a meal in 2015 that would weave together so many strands of my thinking about food, eating, and genetics.

The chefs are husband and wife Oliver Piras and Alessandra Del Favero, a beautiful young couple who work effortlessly, compassionately, and in harmony with each other. They are also perfectly matched in stature and temperament. Alessandra, a classic Italian beauty with fine features (and the olive skin and lustrous dark hair we long for), is the quieter half to Oliver's more rustic, energetic

Sardinian presence. His background includes interning at Noma in Copenhagen (chosen the world's best restaurant four times) and stints in great kitchens all over Europe, including El Celler de Can Roca in Girona, Spain, a Michelin 3-star venue. AGA, which received its first Michelin star in 2016, has been written up in *Saveur* and mentioned in prominent newspapers. (The name AGA, Chef Piras told *The New York Times* is "like *acqua*, or water because we're simple, limpid and transparent.")

In November 2015, a small group of friends decided to try to snag a table at Christmas time. However, as our excitement grew, AGA became not only a *must*, but also a task. How would we fit eight people into a restaurant that has four tables booked well in advance? Lots of correspondence and a few days of "*domani domani*" (tomorrow, tomorrow!) and then that long-awaited jewel of an email: December 23rd, tasting lunch for eight.

We drove from our rooms in the charming alpine ski town of Arabba, through forty-five minutes of winding, narrow, steep scenic passes shrouded by the towering Dolomite peaks down into Cortina, and then on to the tiny village of San Vito di Cadore. We walked in to find that AGA had simple, beautiful, pine-board walls, a serving table at one end of the room, and four diminutive, modern tables with equally sleek, yet simple chairs. The restaurant serves two sittings each day. No fussy folded napkins or tablecloths. One sprig of fragrant juniper in a simple vase as table decor. The restaurant faces east to the mountains, so its many windows allow shafts of sunlight to enter the restaurant at a shallow angle, offering subtle illumination of the otherwise spare space.

That afternoon, the staff presented a tasting menu of eight courses, which somehow turned into thirteen, with Oliver and Alessandra coming out to introduce each one. Both were adorned in timeless chef attire—classic white tunic half-covered by an almost denim blue apron complete with a subtle AGA logo. This is no showplace, and both chefs' aprons bear witness to the constant round of *mise en place* for the day's meals and from the loving—if messy—process of preparing the fermentations and broths that capture local foraging and harvests from the previous months.

Before I take you through that revelatory meal, I want to introduce you to the concept of culinary genomics, so you can understand the impact AGA's meal had on me. In short: Nutrigenomics + Nutrition Science + Food Science + Cooking = Culinary Genomics. Genomics or genomic science enables clinicians and researchers to use information about one's genetic makeup to determine a course of care to create positive health outcomes. You might be tested to see what errors on your genes might, in combination, lead to certain kinds of chronic disease like high blood pressure or diabetes, or prevent them, too. Such outcomes can be mitigated by changes in diet, exercise, stress levels and, in some cases, medication to influence how your genes do their work throughout your life.

The field of *nutrigenomics*, genomics seen through the lens of nutrition science, helps us understand how specific nutrients and other non-nutritional compounds found in food influence gene activity, including how your DNA creates unique recipes for proteins. (This is not to be confused with *nutrigenetics* which explores how individual differences in our genes can, in turn, affect how we absorb and use food and the nutrients it contains.) Culinary genomics merges nutrigenomics, nutrition science and food science (how food responds to growing, storage, and cooking techniques) with the culinary arts to translate these fields of science into precise information about how to choose, prepare and eat food that relates to your genes. You can think about it this way. Nutrigenomics informs us how food interacts with specific genes and in which way (cooked or raw for example), the culinary arts matches ingredient knowledge, preparation skills and cooking techniques to get food on the plate in a way that maximizes the information it provides to the individual's genome.

Now, to the food! As I sat down to enjoy my meal at AGA, it became quickly evident that every course wholly embraced not only the idea of food as medicine, but also mirrored many of the aspects of culinary genomics that we're going to meet – and use – throughout this book. This was immediately clear from two principles that were threaded throughout the menu. First, there was Oliver's dedication to local food, raised in traditional ways that embraced the natural abundance and deep nutrition afforded by local land. Fish is from local streams. Pigeon, hare, and other animal proteins are raised from area farmers dedicated to conserving the art of raising the finest meat in its natural environment. Those happy, well-adapted, well-fed, and active animals consequently offer a complexity of nutrients, particularly omega-3 fatty acids (EPA and DHA in their natural, ready-to-absorb forms) that enable them to immediately interact positively with the body. This bioavailability dynamically helps to reduce an underlying condition of so many diseases: inflammation. Locally sourced fresh cheeses offer the same level of nutrient quality, providing a supporting cast of fresh, immediately available nutrient "information" to support the proteins our genes produce.

Second, Oliver has a deep dedication to the art of fermentation, evidence of which is found throughout his dishes. (After the meal, Oliver would describe the change in his own body and health as he integrated his house-fermented foods into his life and, of course, onto his menu.) Fermented foods are a foundational component of culinary genomics since the gut—and its microbiome—is the gateway to health. A finely tuned gut allows us to absorb the nutrients that serve as information to the body and its complex biochemical pathways. Fermented foods deliver a unique set of information, providing the body with useful information for its 24/7 business.

The fermentation that Oliver works into his menu creations seeds the beneficial bacteria in the gut. We now know that some compounds, such as the sulforaphane which is created when we chop cabbage or bite into Brussels sprouts, can activate influential genes that have a dynamic effect on our health. This same sulfora-phane compound can be produced by beneficial bacteria in the gut. This being the case, bacteria in the gut contributes to food-gene relationships and subsequently, to our health. I'll be talking a lot more about sulforaphane in the Master Ingredients chapter and about gut and the role of bacteria in the Enablers chapter.

So how did Oliver's menu ideas and dedication to local ingredients unfold on the plate? Let's start with a few very simple things in no particular order. First, on the table there is always homemade bread and butter. The flour is sourced from Padova, two hours away, and the bread is produced with an active starter in AGA's tiny kitchen. The active starter initiates a fermentation process which enables the release of a rich diversity of minerals from the flour that would have been otherwise bound up and barely accessible to the body. Butter whipped-up from local cream infuses the nutrient landscape with precious omega-3 fatty acids whose reach, in the Genomic Kitchen, goes far beyond classic "heart health" wisdom to "good for your brain." Add to this more omegas that would appear later in the wild hare ravioli and, once again, the menu is infused not only with flavor from wild food but also with the optimized nutritional advantage that does not exist in factory-raised or industrial-farmed food equivalents.

Next, a lovely velvety risotto of rice grains collapsing into a soothing creamy "broth" was enriched with fermented *umboshi* plum and fragranced with the skin of a local cheese. The dish was dusted with smoked Chinese green tea, where green tea itself is a dominant source of the bioactive molecule catechin, often referred to by its full name ECGC-epigallocatechin gallate. (Bioactives are natural substances with the ability to turn on genes whose proteins help us neutralize damaging free radicals in the body.) ECGC in the form of green tea has been the focus of ongoing research studying its role in supporting the body's antioxidant defense mechanisms in ways that may possibly play a role in cancer mitigation.

Then came gently roasted sea bass served with barely sautéed *rapini* and a bright, almost incandescent *samorigilio* sauce from Sicily. Nutritionally, we know that the deep-green, broccoli-like *rapini* are rich in antioxidants. From a genomic perspective, this delicate vegetable in the powerful Cruciferae family acts like a valve turning on a veritable fire hose in the cell to extinguish pesky and harmful free radicals. The garlic, olive oil, oregano, and parsley of the classic *samorigilio* sauce, all bioactive enhancers as well, accentuate the fire hose effect. Thus, through the lens of genomics, we start to view the food on AGA's menu in a completely different light.

Thirteen courses and at least four hours later we were amazed—everyone by the food—but me, even more, by the overall entirety of the health-promoting ideas that informed the menu.

Chef Oliver was creating and honoring a circle of life, a sustainable cycle, using what was local to the land. I enjoyed it thoroughly as an eater, but the dietitian in me—as we just saw—was inspired by how he combined and prepared ingredients in ways that transformed ingredients into dishes that were not just spectacular but truly medicinal. I discovered that Chef Piras was not only cooking for the soul (ours and his), but unknowingly cooking for our *genes*—the concepts of culinary genomics set right on our plates.

Culinary genomics is a term I devised to describe the union of genomic science, nutrition science, food science and the culinary arts to transform the home kitchen into a resource for modern day health. The discipline involves the selection and preparation of ingredients that are designed to influence key, health-giving aspects of gene behavior. In Genomic Kitchen terms, that means paying particular attention to antioxidants, inflammation, metabolism, and your microbiome. It was incredibly exciting for me to witness this cutting-edge, modern concept of Western medicine in action halfway across the world in a tiny restaurant, nestled snugly at the feet of the majestic Dolomites. The irony, of course, is that to the locals, culinary genomics is nothing more than the way they eat every day.

I wanted to learn more about the inspiration behind Chef Piras' menu, so I asked if I could see the kitchen. It was immaculate—and so tiny it was hard to believe that this was the origin of the grand feast we had just enjoyed. He proudly showed off the back door of the kitchen, which was signed by all of the Michelin-starred chefs who had visited the restaurant. (AGA has a single Michelin star and has for several years.) I wasn't sure how to broach the subject of cuisine and genomics so I asked him directly, "Do you know you're practicing culinary medicine with your food here?" He paused, perhaps considering the random and peculiar nature of my question and said, "Well yes, we use a lot of herbs and spices...."

During our conversation, it became apparent that Piras was aware of some of the health benefits of the ingredients he worked with, particularly the plants and herbs that grew and were foraged locally. He knew the art he was creating was speaking in a unique way to the mind, soul, and the heart. But what *I* noticed was how his food talked to the physiology of each one of his guests. He was unknowingly selecting and preparing ingredients that could have a direct influence on our genes.

As I look around the world, Olivier Piras is not the only chef creating culinary medicine in a restaurant. Understanding the way food interacts with our genes is what I call "the food-gene relationship," and it is at the heart of culinary genomics. Leveraging this interactivity, we can influence how (and which) proteins are made to create balance in our bodies, preventing and fighting off diseases and allowing us to live the very best expression of ourselves. And what I want you to know is that everyone can learn how to do this in their own kitchen.

Genomics uncovers why some people struggle more with weight than others and why some people may be more likely to suffer from anxiety, depression, or struggle to control their blood sugar. By understanding the nature of how food, the environment, and our lifestyle choices influence our genes we can begin to understand health phenomena such as longevity. In this way, genomic medicine cuts through the nutrition noise.

We are at the dawn of a new era in food and a new era in medicine. Never before has the kitchen played a more pivotal role in health. Whether you are a classically trained chef, an avid home cook, or just an individual trying to eat for your well-being, these are watershed times. We have finally reached a moment when we can empirically trace the connections between food, how you prepare it, and how that affects your innate biochemistry. Medicine and the culinary arts are now walking hand-in-hand, forging perhaps the most dynamic and revelatory food conversation and health opportunity in history.

Chapter Two: Eating For a Lifetime

Discovering New Gems for Your Health and Your Plate

In November of 2016, I found myself sitting in a *taverna* in Karyes, a tiny local eaterie in the Peloponnese region of Greece. I had wandered in with my friend, Matina Chronopoulous, a naturopath from Athens. The place was empty. Outside of the tourist season, it really only served the stray visitor or locals on the weekend. It was so innocuous that I would have walked right past it without Matina at my side. As my luck would have it, we happened to enter the restaurant through the kitchen where we first caught a glimpse—and smelled the wonderful aromas—of what we'd soon feast upon.

The visit to this traditional village was part of a brief journey through the region. I was researching the bones of a culinary-cultural-historical tour I was planning that would be grounded in culinary genomics. I create these programs to link food and current nutrition science to specific regions around the world that retain their originality of life via their foodways. The idea is to allow visitors to make an experiential connection between what we now know about specific ingredients and how traditional food and associated lifestyles support our genes and, consequently, our health.

Walking through the *taverna* kitchen that day, I saw chestnuts, persimmons, pomegranates, and heaps of the wild greens ubiquitous to the Greek table—greens which the Greeks simply call *horta*, whose literal translation is "weeds." Sacks of figs, potatoes, and walnuts sat alongside more fresh produce. The dining room was rustic, with simple dark tables, dark wood accents, and a single door with square glass panes leading into the restaurant from the street. That front door had been locked, which was why we had gone around to the side door and then found ourselves in the kitchen. We passed a chef who was sautéing a veritable mountain of those wild greens, a mixture of wild spinach and chicory, fennel leaves, nettles, poppy leaves, dandelions, purslane, amaranth, beetroot leaves, and others I couldn't identify. As we passed, he heaped them up on a plate so that they were swimming in olive oil. He worked with the exuberance of a young chef—but did I mention he was 91? His wife, who cooked alongside him, was 90.

Throughout Greece and particularly here in the Peloponnese, the native diet reflects the season. The late fall season when I was visiting was punctuated with autumnal roots, nuts, fruits, and those wild-foraged greens, a difference from the stone fruit of the summer and the changing citrus varieties for which Greece is renowned. Many of the natives fill their plate at lunch and dinner with foods they foraged that morning, experts in identifying, harvesting, and preparing the flora that surround them. This leaves the markets to us visitors who marvel at the piles of greens that often have no name in English, being native to Peloponnesian soil. Wild greens and wild fruits are a source of pride to the local population and the traditional restaurants serve them up without a trace of a formal menu. Indigenous greens and berries change with each season, giving the regional plate a constant revolution of flavor, texture, and nutritional value (remember Chef Piri's AGA magic menu?) which, unknown to most of the foraging locals, affects their genes and is so linked to their good health. More common vegetables also turn with the season, with classic eggplant, tomatoes, zucchini, and artichokes providing texture from spring until fall.

Back in the *taverna*, we took our seats in the dining room. Instead of a menu, we were greeted by a server who quoted the daily fare from memory, accompanied by some head scratching as he tried to remember everything in each dish. Matina dutifully translated each selection. All the dishes reflected what had been recently picked, foraged, or slaughtered. Most of the ingredients, he told us, had been gathered earlier that day.

We settled on goat meat and chicken, a mix of foraged greens, fried potatoes, and local bread from the village bakery to sop up the freshly pressed olive oil. We weren't counting how many calories were in each dish and the portion sizes seemed random—the greens being the largest. We didn't care. We were eating food from the land, replete with nutrients drawn up from soils that have been sustainably worked largely without chemicals for thousands of years.

The people in this tavern, in this village, and likely in the whole region live without dietary guidance—and they live extraordinarily long lives. They instinctively "eat their view," which naturally changes with every season. Just like AGA's dedication to their local and seasonal food, the Peloponnesian diet is based on the food that can be grown, foraged, or raised immediately around them. Thus, their "guidance" and inspiration come from the land and the season. The demands of working that land and an active lifestyle determine if you pile more on your plate or not. In poor winter weather, farmers are probably not spending so much time out in the fields. Hungry? Eat more. Not hungry? Don't bother to eat. This simple guidance is born of a connectivity to oneself, a place, and perhaps a less complicated (though not always easy) life, lived with respect for the land and for each other.

Why am I so interested in far-flung food, lifestyles, and ways of eating, particularly in the Mediterranean region? Well, my conventional training in nutrition always told me (and I guess still does) that the Mediterranean Diet is *the* diet to follow. The one that avoids so many health problems if we would just eat that way. I grew up in Europe and have eaten, hiked, cycled, and worked my way around the western Mediterranean region. And guess what? Even the food within *one* European country, say Italy, is completely different from one region to another. Try telling the Venetians they eat like the Pugliese! Which leads to the question, what exactly is the Mediterranean Diet given that the Mediterranean covers three continents, each espousing such uniqueness in culture and food traditions? I got a little fed up with the Mediterranean Diet generalities and felt there was and is a different way to understand its attributes and how they're connected to observed longevity, lower incidence of heart disease, and also Alzheimer's disease. So I decided to revisit the Mediterranean Diet again, but this time using my new lens of nutrigenomics. By this I mean using the science of how the food we eat relates or "talks" to our genes as a better way to seek an explanation for the health phenomena associated with this corner of the world.

Let's go back for a moment to the menus at the tiny Greek *taverna* and at AGA in Italy. For sure, the two are worlds apart in terms of their culinary approaches, landscape, and clientele but they're united by their dedication to local and foraged foods on the daily menu. They also share the use of greens and herbs, which appear in abundance on both menus. So the question is: what's in those greens and herbs? A study by researchers at the University of Athens looked at the nutrient composition of greens and herbs commonly foraged and appearing on the Greek plate around the country and its islands. These include Queen Anne's Lace, bitter dock, wild fennel, and chives. While these ingredients checked the box for good sources of the usual mineral players such as sodium or calcium for example, researchers also found them to be a rich source of a variety of bioactives. (Refresher: Bioactives are naturally occurring substances in plants with the ability to turn on or off gene behaviors that help neutralize damaging free radicals in the body and shield the inflammatory process.) When the researchers compared the nutrition composition of commonly available foods across Europe such as lettuce, celery or apple, they found that they also contained the usual mineral content, but the density of bioactives was miniscule in comparison.

In another study for *Clinical Nutrition,* researchers looked at wild greens and herbs, this time on the island of Crete, and noted that the more rural the population, the more likely the diversity of wild greens consumed. "The village residents," the study stated, "eat more often wild greens and herbs (two to four times per week), compared with city residents (one to two times per month). The most often consumed greens are purslane and capers, while the most often consumed herbs are chamomile, sage, and dittany." This last herb is used in "mountain tea," consumed widely across the island and is in the same family as mint, rosemary, basil, sage, marjoram, and oregano. Of particular interest to me, other than perhaps the similar quiet lifestyle shared by the people native to rural Crete and the Peloponnese, is that purslane is a robust source of alpha-linolenic acid, a precursor to healthful omega-3 fatty acids and capers are very rich in quercetin, a highly influential bioactive which we'll learn more about later. Across the Greek islands, those wild herbs and greens studied by the University of Athens—foraged for the home plate and which show up at so many of the local markets—are themselves rich in quercetin. While quercetin does not appear on any food label and may not be familiar to you, genetically speaking it is as ubiquitous as traffic lights in the city and with much the same power—to start, stop, and slow down certain common bodily processes (inflammation, and detoxification, for example) regulated by proteins produced by your genes. Let's investigate a little more by looking at the connection between food, genes, longevity, and the Mediterranean islands of Ikaria and Sardinia.

Longevity, Diet, and the Mediterranean

All of us want to live a long, healthy, disease- and pain-free life. At the end of our life we want to sail off into the sunset with no pain and no regrets. A life well lived. So why do some people live longer than others? Some people struggle with health issues all their lives, while others do not. Surely if we could control our destiny and age painlessly and with contentment this would be the path we would choose. Scientists also ponder these questions, trying to untangle the complex codes that support health rather than induce disease. What we now know through genomic medicine is that certain patterns of genes and their variants can confer protection against age-related diseases, such as cardiovascular disease, Alzheimer's, and cancer. And perhaps no region in the world has been studied as much as the Mediterranean with this in mind. When we can protect ourselves against certain diseases, we, in fact, promote longevity.

Just to be clear, one SNP, or genetic mistake, does not a chronic disease make. Scientists often take a polygenic (meaning more than one gene SNP is involved) versus a monogenic (where only one SNP is the cause of problems) approach to understanding how our gene variants inform disease. Nowhere is this clearer than in Alzheimer's disease. It is the synergy of a number of gene variants that inform the disease process and not a single rogue gene.

Journalist Dan Buettner and his research team have explored the critical relationship between food, health, and longevity as it is associated with people living over the age of 100 in his Blue Zone writings, which began almost twenty years ago. The centenarians his team studied live in what he calls Blue Zones—Costa Rica, Okinawa, Loma Linda in California, the Italian island of Sardinia, and the Greek island of Ikaria. We'll focus on these last two as we explore longevity and diet through eyes of culinary genomics and my interest in examining the Mediterranean Diet through the lens of genomics.

In longevity research, people who live to be over the age of 85 and older are considered *long-lived*. People who live to be over the age of 100 are designated with the term *exceptional longevity*. (There is also a term called *super longevity* for people who live more than 110 years.) In the western world, we continue to witness growth in the early onset of chronic diseases. The earlier the onset, the shorter our lifespan. Paradoxically, we're also seeing growing numbers of centenarians at the upper end of the age spectrum, notably in the Mediterranean region. Why?

Genomically speaking, how the body handles blood sugar, packages fat, or manages blood pressure can be the difference between having a heart attack at age 55 or living to 90 or 110. In general, along with their gentler approach to living, Blue Zone centenarians appear to share an ability to efficiently handle these exact metabolic processes that, if they went awry, could trigger health problems. Efficient management of key metabolic processes allows the body to hum along with a stable rhythm, achieving physiological nirvana, or balance. So it's tempting to say that centenarians are genomically blessed, right?

There is no magic genomic blueprint shared by all centenarians. There is no one perfect set of genes that suggests you will live to be 100 years old or more. That being said, a burgeoning field of research is investigating the role our genes play in longevity. Two frequent research questions are, how does the Mediterranean Diet interact with our genome, and could this interaction explain the lower incidences of chronic disease in the region? A much recognized and cited landmark study, The Predimed Study, investigated these questions. Results illustrated that a Mediterranean-type diet rich in olive oil appears to modulate more genes and "the most prevalent genetic pathways were related to atherosclerosis and hypertension" as compared to a lower fat version of the Mediterranean Diet or a non-Mediterranean diet. Alas, the oft-cited Predimed Study ran into trouble in 2018, generating a veritable scientific brouhaha when an independent group of researchers questioned some of the methodology and data analysis of the work. Subsequent reexamination of the research and summarized by the Harvard School of Public Health seems largely to uphold the study's original conclusions.

Another study analyzing data from the Predimed Study correlated intake of the polyphenol-rich (in other words, bioactive rich) Mediterranean Diet with lower levels of inflammation and reduced risk for cardiovascular incidences among individuals at high risk for cardiovascular disease. Two of the inflammation markers are related to the Nf-kB gene that we'll be talking about a little later. A further study examined the relationship between the Mediterranean Diet and cancer. In this study, researchers reviewed the findings of 28 trials analyzing a total of

570,262 individuals and concluded that the Mediterranean Diet is "associated with reduction in overall cancer rates as well as significantly lower rates of digestive tract cancers." And one more study published online in May 2018, evaluated the diet quality and health risk outcomes of 6,572 Mediterranean men and women with a mean age of 65 years, who regularly consumed fermented dairy products like yogurt and cheese, products often criticized for their higher fat content but nevertheless are associated with a healthy gut. Researchers found that inclusion of these foods positively correlated with better diet quality overall in this cohort, equating to more inclusion of fruit, vegetables and nuts and less intake of alcohol and refined carbohydrates. Moreover, and I guess to be expected, they reported higher HDL-cholesterol (the good kind) and lower incidence of hypertriglyceridemia (elevated total lipids) among those people eating fermented foods. You can learn more about the health-conducive properties of fermented foods and why I include them in the Genomic Kitchen Ingredient Toolbox in Chapters 12 and 13, about Enabler ingredients.

Perhaps the best summary of why the Mediterranean Diet has long been associated with health and related longevity is captured in this sentence from a November 2016 paper by the Department of Preventive Medicine and Public Health for the School of Medicine in Navarre, Spain entitled "Mediterranean diet and life expectancy; beyond olive oil, fruits and vegetables." The authors noted, "In the era of evidence-based medicine, the Mediterranean diet represents the gold standard in preventive medicine, probably due to the harmonic combination of many elements with antioxidant and anti-inflammatory properties, which overwhelm any single nutrient or food item. The whole seems more important than the sum of its parts." In other words, it's about quality and diversity, not just seafood and olive oil.

And yet there is a whole wide world beyond the Mediterranean, lots of healthy food cultures too, many of which we are only just discovering. For example researchers studying global health note the lower incidence of colon, prostate, and breast cancers among Asian cultures that "consume more vegetables, fruits and tea than populations in the Western hemisphere do." Here researchers question whether the bioactives in these foods are instrumental in conferring a protective effect against cancer. From our understanding of how bioactives are implicated in the signaling pathways associated with cancer, the answer could well be yes. If you are interested in the composition of diets from different world cultures and how they have changed in a fifty-year period (1961-2011), visit this fascinating interactive website designed by National Geographic entitled "What the World Eats."

Beyond food, Blue Zone research also indicates that living a low stress life, being naturally active, feeling part of a supportive community, and connecting to spirituality or religion is just as important to healthy longevity as the food you eat. In other words, the lifestyle you provide for your genes is as important as the food you ingest to fuel them. In "Genetics, lifestyle and longevity: Lessons from centenarians," researchers noted that "Heritability estimates of longevity suggest that about a third of the phenotypic variation associated with the trait is attributable to genetic factors, and the rest is influenced by epigenetic and environmental factors." In other words, genes are important to a long and healthy life but their expression is subject to the quality of the environment you provide them. So, for example, if you have a good genomic profile, devoid of gene aberrations that may predispose you to inflammation and oxidative stress and then you habitually expose yourself to toxic chemicals, emotional stress, and poor relationships, your health may suffer regardless of your original genomic gift and sensible food choices. If you are interested in learning more about how environment and social factors impact our health, I encourage you to take a look at the emerging field of research called Human Social Genomics that tracks how the world we live in relates to health outcomes through interaction with our genes.

Now that we've looked at genetics, lifestyle, and longevity, let's get back to food and how to prepare it to optimize your health like those Blue Zones in the Mediterranean and their diet. In a 2012 article for *The New York Times Magazine,* "The Island Where People Forget to Die," author Dan Buettner included the story of Stamatis Moraitis, who was over the age of 90 when the two met. (Watch a clip of Moraitis.) He came from Ikaria, a Greek island in the Aegean Sea, but had moved to the USA. After thirty years in the USA, he was diagnosed in his mid-60s with lung cancer and decided to move back to Ikaria with his wife. He said in the article, "I go home and wait for the death." But a funny thing happened on the way to his deathbed, as he said chuckling. "I went to wait for the death...and no death." He had started to eat pure wild herbs again, planted a garden, consumed his own wine from the family vineyards and used copious amounts of olive oil, famously saying, "I never use butter." Slowly, he began to return to health. As he waited for death, he worked his vines and harvested his grapes and, after thirty years, was still very much alive. He said, "It must be the pure air. The water is beautiful and clear here. You drop a dime in the ocean you can still see it." He noted of course that if you did the same in the ocean around Athens, you would not see the dime "because the water, it is filthy!" In his simple way, Moraitis, who has since passed away, was saying it's about the food, the clean air, the way we live.

Is the Mediterranean Diet an Exaggeration?

Before we look more closely at the food of Ikaria, let's pause to take a brief look at what I call the "Mediterranean Diet Fallacy." This is the notion that every one of the over 35-million Mediterranean inhabitants in 21 countries is a walking testament to the benefits of consuming fish, olive oil, wine, and fresh vegetables; that the diet is replete with beautiful fish and seafood and markets with bountiful produce, all set in this idyllic landscape close to the azure sea. In reality, while many people can see the ocean—many cannot. Even those who can see the ocean may not have the financial means or even transport to get there, and fresh fish is, relatively, expensive. While they may not eat fish three days a week, they are likely eating wild rabbit, venison, goat, pork, and whatever can be raised or grazed locally. Certainly free-range derived animal protein delivers the complement of DHA and EPA that the Mediterranean Diet proponents rage about, but the point is that it is animal-based—not from fish and seafood—and even then likely not to be consumed in huge quantities. (DHA and EPA are essential components of those very healthy omega-3 fatty acids we'll learn more about later.)

The second point about the Mediterranean region is that much of it is quite mountainous, definitely isolated in many parts, and frankly dry and more akin to high desert than tropical or even temperate zones. The media has inculcated us with the idea that the Mediterranean geography is all Provence and Amalfi. To the contrary! Most places are not tourist destinations. They are very simple, with locals eating simple fare (much of it procured locally) and, depending on the region, complementing it with wild foraged flora and fauna. This is particularly evident in the Peloponnese.

The third point is that, yes, folks do drink wine regularly, but not eight-ounce pours of it. Much of the wine is drunk in the home and has been dispensed from the local cooperative usually into a reusable, large plastic container, *sans* fancy label or even glass bottle. They do use olive oil liberally in all forms of cooking at all levels of heat and are neither counting the caloric consequences, nor worried about bathing their food in fat. Most of these people are generally not wealthy either. Those picturesque villages, every stone wall dripping with bright flowers in pots, and those idyllic coastal scenes are symbolic of places that we visitors flock to and not of the citizens of these regions. They live simply, often in the same homes or villages of their childhood, and have known the same people all their lives. Life, for many, is simple as is most of the food in the Mediterranean. While certainly modern life and food is reshaping old traditions and the rhythm of life, in the backcountry of the Mediterranean, some things remain the same, and that usually means keeping it simple.

Stamatis Moraitis lived on the island of Ikaria in the eastern Aegean Sea, close to Turkey. Ikaria is a longevity Blue Zone. The other Mediterranean Blue Zone is the Italian island of Sardinia, lying off the country's west coast. What we see consistently on both islands, Ikaria and Sardinia, is the regular consumption *herbs* in the diet. Many of them are wild and thus foraged when needed. Herbs are also ingested in the form of local teas. Similar to the "mountain tea" consumed regularly across Greece that contains various native herbs and flowers, on the island of Sardinia, milk thistle is the tea of choice.

So why are we drawing attention to the wild greens and herbs of these islands? Because they happen to be a very important key to understanding how to eat for your genes as they all contain large amounts of those substances called bioactives.

What are Bioactives?

Bioactives are non-nutrient dietary components concentrated in specific foods. And they directly influence how genes behave and how they do their work. They are not the same as vitamins (think vitamin C) or minerals like calcium, and they don't have any caloric value either. Relative to gene behavior, bioactives function like a switch that, when activated, sets in motion a series of biochemical steps, akin to knocking over a series of dominos. The end result of this cascade is the activation (and sometimes deactivation) of certain genes. Your genes produce specific proteins whose work is instrumental to the inner workings of your body.

The study of nutrigenomics has revealed which bioactives influence (or modulate) which specific genes. This is important when we think about our health because if we know which genes produce inflammatory molecules, for example, we can then introduce specific ingredients into our diet that contain bioactives to help you turn off or tone down those genes. Ah ha!

One example of a bioactive is quercetin. You'll recall that researchers discovered large amounts of quercetin in the wild greens and herbs commonly found on the daily plate of many Greek people. Quercetin is a bioactive that has been demonstrated through nutrigenomic research to turn off pro-inflammatory genes. Let's connect the dots: Inflammation left uncontrolled causes biochemical imbalances which can result in a variety of adverse health conditions, including heart disease, cancer, Alzheimer's, or premature aging of biological systems. The Mediterranean is famous (in general) for a lower incidence of these conditions as well as for greater longevity. Could it be that centenarians are eating simple foods that have a particular disease-fighting effect on their genes?

Bioactives: A Powerful Biological Tool for Better Health

Let's look a little more deeply at bioactives and then situate them in the context of the islands of Ikaria and Sardinia. To set the stage, consider the idea of a gene by thinking about a general leading a regiment of soldiers on the field. The general represents so-called Master Genes that have the power to activate a number of individual genes. When the Master Gene is activated, orders in the form of genetic information are transformed into proteins. These proteins are like soldiers, each with a specific job to do. Proteins perform jobs like building bone or muscle for example, or playing important roles in supporting the biochemical pathways that permit the body to function, performing tasks like regulating blood pressure or replacing worn out skin cells.

To put this in context, consider the example of how bioactives influence genes that impact the body's inflammatory response. When we look at living a long, healthy, disease-free life, it is important to understand the role of inflammation. Inflammation is the body's natural response to injury or exposure to a foreign substance or as an aid to healing after trauma or surgery. A certain amount of inflammation is required to mount the body's immune defenses to protect the body and guide it towards healing and balance. Things go haywire when we can't turn off the inflammatory response and are left with a low-grade, constant state of inflammation that has a noxious effect on the body.

Inflammation is at the root of much chronic disease and works in a vicious cycle with oxidative stress which means that the more we control it, the healthier we'll feel and the longer we'll live. The more chronically inflamed we are, the sooner we are going to die, literally. Two substances instrumental in turning on the inflammatory response are Nf-kB and TNF-alpha. While both of these act like Master Genes, Nf-kB is actually something called a transcription factor, a sort of "gene helper" that activates the first step in gene expression. You can think of this first step in the same way you start a car by putting a key in the ignition, or nowadays pushing a button! That action fires up the car's cylinders to provide power to the engine. A transcription factor essentially provides the incentive for genes to either get to work, or in some cases, to step aside for a while. Nutrigenomics research indicates that certain bioactives can actually turn off or turn down the biochemical ignition process. We know that one of the ways we can manage inflammation in the body is by including specific bioactives in the diet. One of those bioactives happens to be quercetin. Another is curcumin, a key constituent in turmeric.

Take a look at the islands of Sardinia and Ikaria again and examine some of the native foods that appear frequently on plates there. Take a look at the chart below depicting a selection of foods traditionally and commonly eaten in those regions. The foods highlighted in red are rich in bioactives with known nutrigenomic, or food-gene, interaction and influence. There are some elements common to both diets, such as herbs and wild foods. While chickpeas are a staple on Ikaria, broad beans (fava) are an equivalent on Sardinia. Both promote nutrigenomic activity. As an added benefit, chickpeas and lentils are not only a source of bioactives but also rich in vitamins and minerals. Fennel grows wild on both islands and is eaten as both herb and vegetable. Wild asparagus also grows on both islands. As you already know, the milk thistle herb fuels many cups of tea on the island of

Traditional Foods Consumed on the Islands of Sardinia and Ikaria

with Known Nutrigenomic Food-Gene Connection, or "Talking" Capability

SARDINIA		IKARIA	
Animal Protein	**Legumes**	**Animal Protein**	**Legumes**
CHEESE: FROM SHEEP OR COW MILK	FAVA (BROAD BEANS)	CHICKEN, GOAT, PORK	BLACK EYED PEAS, CHICKPEAS
GOAT, LAMB, PORK, WILD BOAR	CHICKPEAS	**Seafood**	**Herbs**
Seafood	LUPIN BEANS	CUTTLEFISH, COD, HAKE, SARDINES, SQUID, OCTOPUS	ROSEMARY, THYME, OREGANO, SAGE
SEA BASS, BREAM, OCTOPUS, SARDINES, SCAMPI, SQUID, TUNA	LENTILS	**Fruit**	**Other**
	Herbs	APPLES, CAPERS, CHERRIES, FIGS, OLIVES	OLIVE OIL
Other	MILK THISTLE, LAUREL, THYME		HONEY
SNAILS	ROSEMARY	**Vegetables**	MUSHROOMS
Grains (Flour/Barley)	**Other**	CHILI PEPPERS, EGGPLANT, FENNEL, POTATOES, BELL PEPPERS, TOMATOES, ZUCCHINI	PURSLANE
FLATBREAD	OLIVE OIL	WILD GREENS	
SOURDOUGH BREAD	ALMONDS		
Vegetables	FIGS		
FENNEL, TOMATOES, WILD ASPARAGUS	SAFFRON		
	CANNONAU WINE		

Sardinia; while the Greek landscape is lush with wild herbs that are used in cuisine, herbal tinctures, and many teas. And I am going to ask you to make a mental note to come back to this chart later because the animal and seafood groups and, yes, snails, are all rich sources of omega-3 fatty acids, an important part of The Genomic Kitchen's Influencer Ingredient list.

Now let's connect the dots to bioactives themselves. In the figure below, you can see names of specific bioactives that can turn off either the TNF-alpha gene, the Nf-kB transcription factor, or both. While many of these names might be unknown or vague, you might be able to connect gingerol to ginger and cinnamaldehyde to cinnamon. Others, like chrysin and apigenin, might not be so familiar.

Bioactives that Can Interfere with the Action of Nf-kB

Apigenin	Curcumin	Gingerol
Luteolin	Cinnamaldehyde	Quercetin
Caffeic Acid	Ellagic Acid	Resveratrol
Capsaicin	Epigallocatechin Gallate	Silymarin
Chrysin	Genistein	Sulforaphane

Now let's take this a step further by looking at the prevalence of these particular bioactives in foods that are commonly consumed on the Islands of Ikaria and Sardinia. Warning, this is another "ah ha" moment!

What you might notice is that many of the foods that are common to these two islands and their traditional cuisine and diet contain the very bioactives that can block (or "down regulate" as we say in nutrigenomics) the Nf-kB transcription factor that inhibits the inflammatory response. It's as if the people of these islands are taking a natural anti-inflammatory through the food they eat.

The Prevalence of Bioactives that Block Nf-kB in the Traditional Foods of Ikaria and Sardinia

Apigenin
CHAMOMILE (IAKRIA)
ARTICHOKE, FENNEL, CELERY, PARSLEY,
QUEEN ANN'S LACE, OLIVES/OLIVE OIL

Anthocyanins
WILD BERRIES, CHERRIES, SOUR CHERRIES

Caffeic Acid
FENNEL , OLIVES/OLIVE OIL, MARJORAM,
OREGANO, ROSEMARY, SAGE, THYME, PEARS,
GARBANZO BEANS, TOMATO

Capsaicin
CHILI PEPPERS (BOUKOVO) - IKARIA

Chrysin
HONEY (IKARIA)

Curcumin

Cinnamaldehyde

Ellagic Acid
GRAPES, BERRIES (MANY FORAGED
SEASONALLY AND WILD)
EPIGALLOCATECHIN GALLATE

Genistein
FAVA (BROAD BEAN)

Gingerol

Quercetin
BROAD BEANS, DILL, FENNEL, CHIVES,
BROAD LEAF DOCK, QUEEN ANN'S LACE,
(WILD) ONIONS, ,GARLIC, CAPERS, LOVAGE,
OLIVES/OLIVE OIL, ELDERBERRIES, RED
WINE, TOMATO, (LEMON)

Resveratrol
WINE, GRAPES
CONNONAU WINE (SARDINIA)

Silymarin
MILK THISTLE (SARDINIA)

Sulforaphane
INDOLE-3-CARBINOL

CABBAGE, WILD GREENS, (DANDELION),
ASPARAGUS (WILD), PURSLANE, WILD
MUSTARD, ROCKET (ARUGULA)

Remember that inflammation and its enabler, oxidative stress, are central to the biochemical imbalances that cause disease. If we can keep these two troublesome processes in check, then we have a stronger foundation for health and longevity. When we think about the Mediterranean Diet and lifestyle through the lens of nutrigenomics, we start to have a different understanding of how food may be playing a role in the health phenomenon that comes to mind when we think of this region. These days it's no longer enough to point to certain categories of food. What's important to understand is how these foods actually work in the body.

Nutrigenomics—genomics seen through the lens of nutrition—helps us understand how specific nutrients and other non-nutritional compounds found in food influence how genes create proteins and help determine our health. Culinary genomics merges the science of nutrigenomics with the culinary arts in ways that translate this new science to the plate in The Genomic Kitchen. Combining knowledge of how food interacts with specific genes—and in which way (cooked or raw for example)—the culinary arts matches ingredient knowledge and preparation skills with cooking techniques that maximize healthy interaction with your genome.

Medicine has for years been reactive and symptom-driven: you developed high blood pressure and the doctor treated it. Today, modern medicine is preventive, advising us do the things we need to do—avoid salt and fatty foods, drink water, get more exercise—to avoid developing high blood pressure in the first place. The same more actively preventative view characterizes nutrigenomics. Unlike fad diets or today's "super foods," this approach is not about eating to prevent one disease or heal another or lose weight or gain muscle. We are talking about optimizing the body as a whole from the ground up, from the deep inner workings of our cells. Food influences the most fundamental and critical building blocks of our health—our genes. How genes interact with the food we eat and the lifestyle we lead is crucial to optimizing gene expression in order to live a long, healthy life. The enemies of healthy longevity—weight gain, diabetes, heart disease, chronic illnesses of many sorts—all begin with inflammation and oxidative stress. In order to influence the health-stabilizing biological makeup of our cells, we have to look at eating differently over a lifetime. Welcome to The Genomic Kitchen and the world of culinary genomics.

Chapter Three: Culinary Genomics

Taking Genomic Information and Nutrition Science Into The Kitchen

In this chapter I'm going to take you "under the hood" of culinary genomics so that you understand its foundational science. If you feel yourself getting a little lost in the details, jump to the end of this chapter where I introduce you to the Genomic Kitchen Ingredient Toolbox, then dive into the M.I.S.E. chapters in the book. You'll see right away that you can get a food-gene relationship on track with no genomic test in sight. But if you're sticking with me for some science, join me here so I can show you around a bit.

The Art and Science of Culinary Genomics

Culinary genomics describes how to walk the combination of genomic science and nutrition science into the kitchen. This emerging field uniquely combines two fields: one new (genomic science) and one as old as time (cooking). Genomics is what differentiates this approach to culinary thinking from older methods, which might have you looking at recipes for deliciousness but also according to their calories, fat, or concentration of nutrients. While nutritional aspects are still important, culinary genomics expands our understanding about those nutrients as well as the ingredients that contain them and how to prepare them for optimal integration by our genes. It's a new view on what we do in the kitchen through the lens of communicating with our body systems through our genes, with food.

Once we understand how genes work, we learn which biochemical pathways are influenced by the proteins certain genes produce and the nutrients required to support those pathways. Then we look at which ingredients are the best sources of the bioactives and nutrients to support the genes and their respective pathways. We also consider how to prepare those ingredients to ensure their precious nutritional components are in their most useful, biologically available form for the body

Culinary genomics is deeply seated in evidence-based research and represents an exciting era in science, and, more importantly for your personal health, in the kitchen. It is a field that will enable you to prepare delicious food at the cutting-edge (pardon the pun) of healing.

Before we can take a deep dive into culinary genomics, we must first understand some basic components of the science behind the discipline that will allow for our relationship with food to make a shift. These four components are:

1. What genes do;
2. How genes function;
3. What happens when genes don't function properly; and
4. How food influences genes and their functioning.

In the following chapters, I will then help you expand your food knowledge in a way that will enable you to stock your kitchen, prepare healthful foods, and fill your plate with ingredients that are the best source of information your genes need. And trust me, your genes will thank you. So will your health.

Spice Cabinet as Medicine Cabinet

Turmeric is a spice that you might be familiar with. It's the brilliant-hued, yellow-orange powder that gives some Indian and Asian food its bright color. You will find it at your store either as a dried spice, or, if you are fortunate, in its whole-root form. You've probably read that turmeric is a powerful anti-inflammatory, but do you know how it works to combat inflammation? The secret lies in its bioactive constituent, curcumin. This inherent compound can turn on a powerful transcription factor, Nrf2, which you'll be reading about shortly. Nrf2 switches on the body's ability to squelch free radicals. Free radicals are atoms with unpaired electrons. Since electrons like to be paired, these damaged atoms search the body trying to steal other electrons, a process, which, if left untended, can cause inflammation by doing a lot of damage and destruction to your cells and tissues. Turmeric has become a potent anti-inflammatory ingredient in the kitchen because you can add it to so many things. Its easy enough to grate some fresh turmeric into your smoothie or a teaspoon of dried works too. Build your own curry powder from scratch, or grab the store version and you can add some anti-inflammatory "medicine" to delicious curries and dahls.

What Genes Do: Genes Make Proteins, the Building Blocks of Our Bodies

Although many of us think of genes as a "you have the genes you're born with and that's that" concept, genes are actually much more dynamic. They don't just determine that you will have black hair and blue eyes, but also that your blood pressure will go up and down; your body weight will likely increase without proper diet and exercise; and that the cells of your leg muscles will let you run that 5K on Saturday as sufficient energy-release is regulated by, of course, those genes. Genes are our blueprint for life. Genes make the amino acids that are the building blocks of proteins. Some proteins become muscle, some become hormones, or enzymes or bone, for example. Proteins build, regulate, and maintain the human body. They are fundamental to life, which is why they are so important.

Here's an example of how genes work. Many of us are familiar with diabetes, particularly Type II or adult onset diabetes. A hallmark of this form of diabetes is hyperglycemia, which you may also know as elevated levels of blood glucose. Insulin is essential to how the body regulates blood sugar. Since insulin needs zinc to do its job, one protein that ensures insulin a supply of zinc is the Zinc Transporter Protein, or ZTP. We know very precisely that SLC30A8 is the gene that encodes ZTP, which is to say that the SLC30A8 gene activates the production of ZTP. Sometimes SLC30A8 incurs production errors which result in a faulty encoding of ZTP. Yes, these are the "spelling errors" I mentioned earlier called Single Nucleotide Polymorphisms, or SNPs. If our ZTP is impaired, it follows that our insulin's effectiveness might also be impaired, because insulin will not get the steady supply of zinc it needs. The end result is potentially abnormal blood sugars. Here you have an example of a protein whose work impacts a hormone essential to blood sugar regulation.

How can we create a work-around, particularly if you don't know, absent testing, that you have a SNP on the gene that makes ZTP? Knowing that zinc is important to how insulin "performs," the first thing you can do is ensure you are eating foods rich in zinc. Animal proteins and seafood provide the richest sources. If those foods don't appeal to you, then up your intake of nuts, legumes, and whole grains.

How Genes Function

Genes are made up of deoxyribonucleic acid, or DNA. DNA is the hereditary material in humans and determines who you are. It defines your uniqueness. Think of DNA as your software for life. Computers have a code called the binary code which uses the digits 0 and 1 in endless combinations to make up programs that accomplish countless tasks. DNA also has its own chemical code of four base proteins, called nucleotides: adenine (A), guanine (G), cytosine (C), and thymine (T). In DNA code, the order, or sequence, of AGCT determines how and which proteins are built.

In computers, the binary code is organized into bits to make bytes. A bit is a digital *pair* comprised of 1s and 0s. Four pairs, or bits, of 1s and 0s are strung together to make a *byte*. A byte therefore comprises 8 digits, or four pairs. While there are variations and other uses, a byte contains the data that can be translated into a singular number, symbol, or a character in text. For example, this byte—01000011—translates into the character "C" using the American Standard Code for Information Interchange (ASCII). Changing or transposing a single digit in the example above—that "spelling mistake"—completely changes the value of the byte and the corresponding number, character, or symbol. So, changing the third digit from a 0 to a 1, the byte now contains data representing the character "G:" 01100111. Thus, a micro spelling error in a bit, for example, can change a number/character/symbol and drastically alter intended results. Just ponder the sentence, "Let's get engaged." changing the first "g" in "engaged" to a "c."

The DNA equivalent of the binary byte is called a *codon*. A codon consists of three nucleotides or building blocks. The human body can choose from four nucleotides—cytosine (C), thymine (T), adenine (A), or guanine (G)—to build a codon. As you can imagine, there are lots of ways you can arrange these letters or nucleotides. One codon provides the instructions for one amino acid, many of which are strung together to make a protein, much like a string of pearls.

Similar to the binary code example above, if one letter (nucleotide) anywhere in a codon recipe is changed or transposed, it changes the type of amino acid produced. Your body will still string together the amino acids to make a protein, but the recipe for the protein is altered. For instance, if the SNP changed a nucleotide in the recipe for hemoglobin, we might end up with the protein equivalent of "hemoBlobin" instead. Hemoglobin is essential for carrying oxygen around the body. Can one protein with a spelling error still do its job? For the most part yes and we'll talk more about that in just a minute.

We call this change in sequence, or spelling error, a SNP which is shorthand for Single Nucleotide Polymorphism. This is how you and I differ from each other. As humans, our genes are about 99.9 percent identical. One of the secrets to how you and I differ lies in the little recipe variations for protein which are accounted for by the 0.1 percent. Yes indeed, that 0.1 percent variation does, in fact, mean that your box of protein recipes is different than mine.

When Genes Don't Function Properly

Think of this recipe variation like this: just as you can type an entire page of text and make simple spelling or grammatical errors, so can I. The difference is where you make your errors versus where I make mine. The text is still comprehensible, but the errors define the difference between your version and mine. Before you imagine the worst, SNPs are a common byproduct of our DNA's work. There are an estimated 10 million SNPs occurring across the three billion genes of the human genome. We all have them, but we all have a different SNP blueprint. One SNP or spelling error will neither build your health house nor burn it down. Instead, it is the pattern of SNPs, and how they relate to each other, that can positively or negatively impact your health.

Clinicians practicing genomic medicine look for patterns of SNPs within biochemical or metabolic pathways. The ZTP example I gave you is *one* example of how a SNP or spelling error in the makeup of this protein (encoded by gene SLC30A8) might affect insulin production and blood sugar regulation. A clinician would never conclude that you might have a blood sugar management problem based on that one SNP. Instead, trained clinicians look at the big picture of where SNPs are; how they relate to each other within one biochemical pathway or between pathways; and what impact these collective SNPs have on how your body functions. Once a clear picture is formed, additional targeted blood work or other relevant tests can be ordered to evaluate the impact of SNPs on your biochemistry. For example, blood work tells us whether a gene producing an enzyme that converts beta-carotene in plants to the active form of Vitamin A (retinoic acid) is producing a faulty protein or working just fine. Genomic information says perhaps that gene is faulty, or has a SNP. Your blood work comes back and the enzyme is working perfectly. Another example shows us a pattern of SNPs that is impacting how your body is handling carbohydrates. We order a fasting blood sugar and fasting insulin and see that both are elevated. This pattern tells us that we need to intervene with diet and exercise to create a workaround for those pesky genes. As one of my mentors always says, genomic information is *one* of the tools we can use in our clinical toolbox. Like a signpost, it provides some direction as to where we might look for solutions.

The good news is that, depending on the SNP or SNP patterns, we can create an effective game plan using food, exercise, lifestyle modification, medication (sometimes), and additional treatment modalities such as massage or acupuncture to influence gene behavior directly, or to create work arounds when genes are "sluggish."

Yes, But I Don't Know My SNPs and I Don't Have Genomic Information

We know that gene SNPs and patterns of SNPs can impact how your body functions. We also know that specific nutritional and lifestyle recommendations can mitigate, or offset, the effects of gene SNPs on your health. Thus, food is critical to how we can improve your health and wellbeing, especially when we know your SNPs. But what if you, like most people, don't know your SNPs? How are culinary genomics and the Ingredient Toolbox I'm unpacking in this book relevant to you?

A fundamental concept to remember is, as human beings we are "built" with a set of genes that are pre-programmed to use food as a primary information source. The role that each gene plays in human health is fundamentally the same for all of us. It follows then, if we know how certain genes function in the body, and which food they need to function, we can create a toolbox of ingredients to support those genes and their functions. As humans, we are not separated by food, but by how much of each specific nutrient we need from food. In other words, while genomic science provides the supporting foundation for the principles I'm sharing with you, the principles can be applied and the ingredients purchased and prepared without ever investing in a genomic test. Let's look at how we can apply these broader principles of how food influences your genes by delving a little deeper into the fundamentals of the food-gene relationship.

Chipping Away at Genomics

You can probably tell that I am a huge advocate of personalized genomics, but all of this comes with a reality check. The Human Genome Project was completed in 2003 and has shaped the current revolution in personalized medicine to include genomics. When new science enters the clinical space, it takes time for health and medical experts to catch up. Regardless of whether you are seeing an MD, a naturopath, a dietitian, or other health expert, for these individuals to learn about and integrate a new field of scientific thinking takes time away from seeing patients. Genomics is not an easy science. It requires a deep understanding of how genes function (genomics); of the function of those proteins that they produce (nutritional biochemistry); and how to evaluate an individual's gene SNPs using laboratory data. Most clinicians practicing today were not taught genomics, so they are learning how to interpret this new science right now. If you are disappointed when your physician or dietitian cannot immediately interpret your newly acquired genomic report, know that many are on a high-speed learning curve, *right now*. If you are interested in personalized genomics, be sure you can identify a clinician who has had training in genomics *before* you order a genomic test with them. This way, as an individual, you know you are working with someone who has the training to help you.

How Food Influences Genes and Their Functioning

Food is not the only medium that influences how your genes behave, but it is certainly one of the most important. Remember that food is essentially raw material for the body. It provides the basic information that your body interprets to operate. You probably have heard food discussed as proteins, fats, carbohydrates, vitamins, and minerals. All these components feed into the food-gene relationship. But there is another critical component we find in food. Remember those bioactives we learned about earlier?

Bioactives: The Key to the Relationship

The Office of Dietary Supplements at the National Institute of Health defines bioactives as "constituents in foods other than those needed to meet basic human nutritional needs, which are responsible for changes in health status." As we learned earlier, bioactives are not nutrients. They are not calcium, zinc, or vitamin A. They are another molecule that doesn't have a caloric value and doesn't act like a nutrient, but does have is a proven influence on health.

As we already discussed in Chapter 2, bioactives can influence how your genes behave or express themselves to produce the proteins we need. Once those proteins are produced, vitamins and minerals support their function in your body, enhancing or multiplying their effect. It's a one-two punch. Remember, bioactives influence gene behavior by sending a signal to your cells, which sets in motion a series of complex biochemical steps. Those steps culminate in one of your genes creating its specific recipe for a protein, a process called gene transcription.

Which Foods Contain Bioactives?

In *The Genomic Kitchen* we are focused on which bioactives have been demonstrated through peer-reviewed research studies to influence gene behavior in humans. While many have been studied for their overall contributions to the food-gene relationship, a much smaller number have been shown to exhibit demonstrated positive effects in human-cell cultures and humans themselves. Navigating the science to narrow down this category of bioactives is a monumental task and not easily undertaken. This is where integrity in the scientific literature is so important.

An example of a bioactive that influences genes in humans is quercetin, which we discussed previously when we talked about the wild greens and herbs common to Greece and its islands. Beyond those foods, we find quercetin in many other common foods, particularly in high concentrations in the allium family: onions, garlic, leeks, and shallots. Also in capers, lovage, juniper berries, radishes, radish leaves, and elderberries. We'll learn more about quercetin and its effect on human health—everything from inflammation to allergies—in Chapter 5 when we discuss Master Ingredients.

While bioactives are critical to the food-gene relationship, they are just one part of it. We know that genes produce proteins, but proteins are ineffective by themselves—they need a robust presence of nutrients, vitamins, and minerals to do their work. Consider a diet that consists solely of fruit. It's obviously not going to sustain your physical needs. Your body needs a variety of nutrients to function. Similarly, genes produce proteins that are essential to how our biochemistry works but they need a large supporting cast of vitamin and mineral "cofactors" to complete the task. We'll talk more about the importance of vitamins and minerals in the Super Foods chapter later on.

Native Plants and Cuisine That Might Just Influence Your Genes

The flora and fauna of landscapes around the world have been part of folklore, medicine—and the dining table—since the beginning of time. People have always created teas and tinctures, made up poultices and added local and regional foods to everyday dishes as part of their place-based traditions and natural medicine. Now, in this era of nutrigenomics, we are gleaning a better understanding of how *some* of these innocently foraged plants may be talking to our genes and acting as a form of medicine.

A study published in 2014 reported on the traditional food knowledge of 39 local elderly people in the Emilia-Romagna region of Italy (think Bologna and Parma). They recorded the common and folk names of local foods, the parts of the plant used, the culinary preparation, and the medicinal usage. The foods included greens, fruits, and semi-wild plants. The participants cited a wide variety of plants, with many belonging to the Lamiaceae and Asteraceae family as well as Rosaceae. The Lamiaceae family includes herbs such as rosemary, mint, and thyme, which have demonstrated nutrigenomic, or food-gene effects. The Rosaceae family includes many wild fruits and berries, some of which we recognize as common fruit, and many we don't. Science has yet to catch up with the full range of their bioactives and their contributions to the food-gene relationship.

Another **study** investigated the use of edible wild plants by people in a village in Western Anatolia in the Aegean or eastern region of Turkey. Here, people collect the plants for their own consumption, but also importantly for income, with many varieties showing up in the local markets. Over 90 species were recorded in the study, with many being consumed once or twice a week, fresh. Among the plants frequently collected were wild mustard, wild radish, and fennel, each of which has known bioactive capability. A further **study** conducted in same region reported consumption of a wide variety of plants belonging to 26 different families, eaten mostly in the form of salads, therefore fresh. The most widely consumed plants belong to the Asteracea family, which includes the daisy but also **lettuce**, **sunflower seeds**, and **artichokes**. Boraganinaceae plants were the next most commonly consumed and included borage. Then came the Apiaceae family also known as the Umbelliferae--including carrots, cumin, dill, fennel, and parsley; the latter having demonstrated nutrigenomic effect.

All of this makes me agree with the wisdom of our ancestors and the rural dwellers of the world, how deeply connected to the food they eat with the knowledge that different species do seem to heal. It's interesting that the further the distance between ourselves and where our food grows, the sicker we become and the more we seem to rely on modern medical intervention to heal us. And yet, for hundreds of years, folks gathered their medicine and healed themselves, give or take a few unfortunate experiments that probably did not end well!

Cooking With Food that Influences Your Genes

The premise of culinary genomics is simple: use very specific ingredients prepared in a certain way to influence your genes and optimize your health. Cooking is the bridge between food and nutrition science. Your kitchen is a space where you can have the deepest influence on the critical food-gene relationship we've been examining. By selecting specific ingredients and paying attention to their preparation, you're starting an ongoing food relationship that will affect your health for the rest of your life. This is a relationship you can nurture and feed with the squeeze of a juice press, the sizzle of a stir-fry, the blade of a knife. And all this without any reference to a "diet."

The concept of culinary genomics is not restrictive. Instead, it opens up the possibilities for building a new relationship to food with strategies that help you eat for health over a lifetime. Along the way you may discover, as I have, that a culture built around good food can make for a life well lived.

Cooking with Food that Talks to Your Genes

A nutritional biochemist mentioned to me a few years ago: raw tomatoes are a great source of vitamin C. Cooked tomatoes are a great source of lycopene. In other words, whether your tomatoes are cooked or raw, they are nutritionally beneficial for you. But in culinary genomics, we add another layer of thinking, namely how to prepare food in a way that best influences our genes. It turns out that lycopene in tomatoes has more influence on your genes when the tomatoes have been either cooked or slowly dehydrated (think sun-dried) than if eaten raw. The reason why has to do with something called *isomers*. An isomer is a compound that has an identical chemical formula, but exists in different versions. In tomatoes, isomers exist in two forms: *cis* and *trans*. The cis form,

as found in cooked tomatoes, is the one associated with nutrigenomic activity, i.e., that can influence our genes. The trans form
cannot. Specifically, lycopene in cooked tomatoes can talk to our
Nrf2 transcription factor, which you met earlier but which we'll be
learning much more about later.

But what if you don't like cooked tomatoes, are there other
sources of lycopene in its cis nutrigenomic form? Why yes! If
you love watermelon, then eat away because it is rich in the cis
form of lycopene. *Bon appétit*! Oh, and please enjoy those height
of summer heirloom tomatoes, completely in the raw. There are
some food rules that simply need to be disregarded when flavor
comes into play!

The Genomic Kitchen Ingredient Toolbox

All the ingredients that influence our genes are included in what
we call the Genomic Kitchen Ingredient Toolbox. The ingredients
in our toolbox are grouped together in categories according to what
function they perform in the body, and in a neat shorthand they
spell out MISE. Mise in a restaurant kitchen is short for *mise en
place*, the French culinary term that refers to the set up required—
every ingredient, pan, utensil, and every element of prep work that
cooks must have in place before they prepare a dish.

- M: Master Ingredients
- I: Influencer ingredients
- S: Super Foods
- E: Enablers

Each category makes an important contribution to the food
gene relationship. Genomically, we are looking at the entire
system where no one ingredient category is more important than
another; but instead, all work in harmony with each other. In the

following chapters, we'll explore the science and function of each ingredient category in relationship to our health, and simple ways to shop, cook, and eat with these ingredients.

And One Final Lingering Question That Always Comes Up: What About Supplements?

When we are talking food and health, the issue of taking supplements inevitably comes up. While we focus on whole foods in the Genomic Kitchen, it is important to mention that for some individuals, supplements can provide helpful adjunct nutrition support. Genomic information can initially tell a clinician which SNPs and collections of SNPs might impact how well a gene in your body responds to food, meaning whether a bioactive or nutrient will trigger a gene to move into action, or turn it off. This is personalized nutrigenomics. Your individual genomic information can also provide clues into whether a gene might produce a protein that impairs how a nutrient is absorbed and utilized by your body. This is a slightly different nuance and is called nutrigenetics. In some cases, ingestion of foods containing the critical bioactives and nutrients we need does not provide sufficient "nutrient power" to work around the impact of those inherent SNPs. Clinicians know this by testing to measure certain biomarkers (or specific lab values) prior to suggesting specific nutrient changes in your diet. Assuming those lab values were not optimal prior to the dietary changes, we look for them to change after dietary intervention. If the change in the lab is insufficient, it suggests the body needs more nutrient "ammunition." The best way to get this is, again, via food and often with supplement support.

This is a perfect example of the power of individualized genomic information. Genomics takes the guesswork out of whether you need a supplement, which supplement you actually need, and how much of that supplement you uniquely need. And I can tell you from personal experience, individualized genomic information can also save you money by getting rid of supplements you don't need, are taking too much of, or, more importantly, that are not actually working for you. Know this. Whole food, in the form designed by nature, is always the best information source for your body. Here nutrients are packaged in a ratio determined by nature and one that your body recognizes. Food first, supplements second and which supplement and how much, preferentially determined by lab testing and not by guessing.

But as my mentors have reminded me over and over, and I now repeat to you, genomic information is one of many tools we can use to build a roadmap for our health. It is accurate and deeply personal, but we cannot rebuild your health through genomics alone.

Chapter Four: The Genomic Kitchen

Reimagining Your Kitchen to Make Food That Influences Your Genes

In earlier chapters we discovered some broad concepts behind eating to influence gene behavior. We looked at different cultures and regional eating habits that embody these conclusions. In this chapter, we are going to get granular. I invite you to step into my kitchen. Well, not *literally*, but into an imagined kitchen in which we look at everything from the utensils on the counters and in the drawers, to the contents of the refrigerator and pantry; in short, all essential elements in the Genomic Kitchen. You will probably quickly identify a lot of familiar ingredients and at the same time realize you don't need a kitchen rebuild or a new set of fancy utensils, either. The Genomic Kitchen is open to everyone.

Why the Genomic Kitchen Is Different

As you recall, genomics is one part of a new era in personalized medicine, where clinicians with expertise in genomics provide solutions based on your unique dietary needs, not mass-market advice and solutions. But, as I have said before, regardless of whether you have a genomic test or not, I want to reassure you that you can easily apply the principles of genomic science in your kitchen and life right now. Our genes respond to the information from food in the same way. The differences between us, and what we can delineate at a more personal level from genomic tests, are small: how much of a certain ingredient or nutrient we might need, *not* whether we need them. Our bodies require a constant flow of consistent food information to operate efficiently. Culinary genomics tells us which food information and how to prepare it so that it is in the best form for our bodies.

So, culinary genomics shape recipes and meal ideas that use specific ingredients prepared in certain ways to facilitate the best environment for important and influential genes to operate in, supporting the biochemical systems that are essential to your health. The Genomic Kitchen introduces you to a completely new way of thinking about food choices, food shopping, and food preparation. I ditch the idea of choosing ingredients that are good for a body part, such as your bones or your brains. I replace that idea with ingredients you can get at your local store that collectively provide foundational support for your body from your cells up. In short, I'm helping you write a new nutrition prescription for your health and engaging you in a brand new food conversation. All that being said, what does this Genomic Kitchen actually look like?

Let's lay out a couple of facts before we go through the kitchen door. First, a Genomic Kitchen is not a kitchen where we create generic "special diets." Rather, it's where we reach naturally for foods that coax your influential genes into action, or support them when they are struggling to, say, tamp down inflammation. Second, it's a kitchen where the food we purchase and prepare supports not only your most influential genes, but also the day-to-day business of our biochemistries. Happily, there are no Nutrition Facts Panels guiding your food purchases in the Genomic Kitchen because the percentages and mathematics and other gobbledygook on food labels are not the language our genes speak. Learn the basic principles of culinary genomics, however, and you can apply this framework to how you eat daily.

Third, while there are a lot of foods we do include in the Genomic Kitchen, there are quite a few that we don't include because they are *not* a fit for many of your genes (or your neighbor's either). Frankly, they are not conducive to health, regardless of marketing hype that might try to persuade you otherwise. In the presence of trans fats, copious amounts of sugar, and unidentifiable "food products" that don't exist in nature (Michael Pollan famously named these "food-like substances"), your health-friendly genes do not respond well, as you will soon see. They in effect become shrinking violets.

What Does This Genomic Kitchen Look Like?

Some of you may have the impression that a Genomic Kitchen looks like a laboratory because we've talked about so much science and it sounds very "scientific." But no, you don't walk into the Genomic Kitchen with a DNA test report, insert it into a machine, and blindly follow the diet it spits out. What we focus on in this kitchen is how specific ingredients work in influential ways in your body. Choosing the right food and preparing it properly can have a dynamic impact on specific genes and biochemical pathways in your body, and subsequently improve your health.

Hear me clearly, the Genomic Kitchen does not require a designer. Nor do you need to demolish your current island and countertops! The Genomic Kitchen focuses on ingredients and their preparation and not on special equipment, elaborate cookware, "smart" refrigerators, or pull-out, built-in pantries. As long as you have the basics, such as a pot to cook in, a knife to cut with, a board to cut on, and a few other useful tools we'll talk about, you can move right into a food-gene relationship in your very own kitchen. In short, you can stand in your current kitchen and start putting principles into practice right away.

Let's Get to the Food

Stocking our kitchen does not require an overnight makeover. Instead, I'll take some time over the next few chapters to help you to understand why we use specific ingredients in our kitchen and don't use others. No need to rush to your refrigerator and pantry and dump out everything in them. To the contrary, you can start to slowly *add* these specific ingredients to your shopping basket at the grocery or your local farmers market and begin to experiment. Also, while I focus on specific ingredients that form the core of the Genomic Kitchen, these are not mutually exclusive of what you have on hand. Think of them as foundation ingredients, but not the only ingredients in your kitchen. In a chopped salad, for example, yes, reach for the fennel on our master list of ingredients—but throw in the dried cherries you love, too. Enjoy a variety of different ingredients to make the meals you want to eat. In the next chapters, we'll explore the "whys" behind these lists below.

What Ingredients Do I Need to Start Stocking My Kitchen?

If you are like me, you want the answers fast! Well, here is the base list of all relevant ingredients across all categories (which I will repeat again in Chapter 9). I include them here as a "first look" for you. In the ensuing MISE chapters, I will slowly unpack them so that you understand why they are in the Genomic Kitchen.

Here they are.

MASTER INGREDIENTS | **INFLUENCER INGREDIENTS** | **SUPER FOOD INGREDIENTS** | **ENABLER INGREDIENTS**

Fruit
Apples
Blackberries
Blueberries
Elderberries
Grapes
Pink Grapefruit
Pomegranate
Raspberries
Strawberries
(Cooked) Tomatoes
Watermelon

Vegetables
Crucifers:
Arugula
Bok Choy
Broccoli
Brussel Sprouts
Cabbage
Cauliflower
Collard Greens
Dandelion Greens
Kale
Kohlrabi
Mizuna
Radishes
Radish Leaves
Rutabaga
Tatsoi
Turnips
Turnip Greens
Wasabi
Watercress
Other:
Carrots
Fennel
Alliums:
Garlic
Leeks
Onions
Shallots

Herbs
Basil
Chives
Cilantro
Dill
Lavender
Lovage
Oregano
Parsley
Rosemary
Sage
Thyme

Legumes
Broad Beans
 (Fava/Pigeon Peas)
Boiled Peanuts

Spices
Ginger
Hot Chili Peppers
Turmeric

Other
Capers
Green Tea
Honey
Olive Oil
Red Wine

Fruit
Avocados
Mango
Oranges
Sun Dried Tomatoes

Vegetables
Beets
Corn
Peas
Peppers
Spinach

Seafood
Albacore Tuna
Anchovy
Herring
Mackerel
Mussels
Oysters
Pacific Halibut
Salmon
Sardines

Animal Protein
Cheese (Preferably Aged)
Meat
Poultry
Whole Eggs

Nuts & Seeds
Chia
Brazil Nuts
Flaxseed*
Hemp Seed
Sesame Seed*
Sunflower Seed*
Walnuts

Legumes
Black-Eyed Peas
Soybeans*

Other
Brewer's Yeast
Mushrooms (Shitake)
Nutritional Yeast
Quinoa
Rye
Sea Vegetables*
Wheatgerm

Legumes
Chickpeas
Lentils
Soybeans*

Seeds
Flaxseed*
Sesame Seed*
Sunflower Seed*

Other
Sea Vegetables*

Prebiotics
Artichoke
Asparagus
Banana
Burdock Root
Chicory Root
Dandelion Root
Garlic
Jicama
Leeks, Onions (Emphasized
 Here As Prebiotics)
Whole Wheat

Fermented/ Cultured
Fish Sauce
Miso
Soy Sauce
Tempeh

Dairy Fermented/ Cultured
Kefir
Yogurt

Non-Dairy Fermented
Sauerkraut
Kombucha

*Ingredients with an asterisk
appear in more than one
category, reflecting different
biochemical functionalities of
these ingredients

Ingredient Nitty Gritty and Hints

PRODUCE

In the Genomic Kitchen, we work with many items you'll find in the produce department and spice aisle in your local market. This is because fruit, vegetables, herbs, and spices are not only nutrient powerhouses, but are also a vital source of bioactives that initiate engagement with your genes. Yes, you want to continue to choose and eat a wide variety of produce and, yes, in lots of different colors. In the Genomic Kitchen, however, you will see that we have a more specific list of foods to focus on—and I'll tell you why as we go along in the book.

Let's begin with the basics. You want to purchase quality produce and then nurture that investment. By quality, I mean not only going for fresh and seasonal produce but also being choosy about the origins of what you buy. Farmers markets, Community Supported Farm shares, roadside stands—these are great because you can get to know the growers and trust them. When I don't know the grower, I choose to buy either organically labeled produce or from a grower whose growing methods I am familiar with. I have learned that these conversations are vital. Food that is not labelled organic isn't necessarily raised with synthetic inputs. Nor does the lack of an organic label mean it's grown without care and attention to flavor and texture. I often buy from many farmers at market stalls that are not embellished with "organic." I buy after I have had a conversation. Often times, I have cycled past their farms, or know through word-of-mouth about their growing practices. And if I am not buying direct from the grower, then I quiz the staff in the produce department about the origins of what they are selling. The consumer-to-grower connection is vital to understand how our food is grown and labeled.

All this being said, how food is grown *does* influence how our genes behave, for example, when additives like fertilizers and pesticides that accumulate in food end up in our bodies. Talk to clinicians working with individuals who have complex health issues, and so often, laboratory results indicate accumulation of toxins, some of which come from that processed and industrially grown food. Toxins have a way of accumulating in the body and can impede normal metabolic processes. While the body has effective mechanisms to remove them, some of us have SNPs in our detoxification pathways which hinder our ability to effectively remove these substances. This is all the more reason to be mindful of what you eat and to understand the "why" behind it. You can support your body's innate ability to detox by preventing the burden in the first place by buying your food carefully—like choosing produce that has been raised with minimal to no chemical inputs. Know thy grower, and if you don't, ask a lot of questions of the store where you are buying your groceries.

Witnessing the Effects of Toxins in Clinical Practice: Mercury Build up in Patients Who "Eat Healthy"

Mercury is a toxic contaminant that can cause a variety of seemingly unrelated symptoms. One stark illustration, much studied, was the industrial pollution of Minamata Bay in Japan in the 1950s and '60s that caused lifelong health injury to thousands of people, opening a research window into the toxic effects of this heavy metal.

Mercury is frequently ingested through the aquatic food chain (in the form of methylmercury), but is also a global pollutant often distributed through water and soil pollutants and dirty air emissions. In "hotspots" around the world, clinicians have linked both industrial and food contamination of mercury to health injury. Clinically, we often see elevated mercury levels among patients reporting a variety of seemingly unrelated symptoms. Here's why. One body of research suggests that mercury depletes a critical mineral, selenium. To detoxify, or remove mercury from the body, selenium works in conjunction with glutathione and an enzyme you will be learning about, glutathione peroxidase (GPX). The glutathione/glutathione-peroxidase system binds and removes toxic heavy metals from the body, which include mercury. But mercury is a nasty molecule and very adept at binding up selenoproteins, essentially making them unavailable for work in the body.

Selenoproteins are proteins which include selenium, or more precisely a selenocysteine molecule, in their structure. Selenoproteins are integral to many systems, including thyroid metabolism, recycling vitamin C (a key antioxidant in the body), and protecting against DNA damage. The good news is that we can potentially overcome the nastiness of mercury, first by avoiding it as best we can, but second by ensuring that our diet is replete with selenium. The bad news is that our diets are notoriously deficient in selenium, which means that we are more susceptible to mercury toxicity. This presents a public health conundrum since seafood, a good source of essential omega-3 fatty acids that we want you to eat, is also increasingly and unfortunately a source of heavy metals as well.

So what's a person to do? If you do consume seafood, familiarize yourself with seafood containing the lowest amount of mercury. An excellent resource is the **Natural Seafood Defense Council's Wallet Card** on mercury levels in seafood. Second, amp up selenium intake in your diet. While some meat, poultry, and egg are sources of selenium, the concentration of selenium truly reflects the diet the animals have ingested. But the amount of selenium in these proteins is quite small in comparison to Brazil nuts, which you should consider as the premium source and add to your diet immediately, if you can. Brazil nuts are in the Genomic Kitchen Ingredient Toolbox. See, we have you covered!

But let me also point out to you that while I have discussed seafood here, mercury is one of several heavy metals that can make their way into our body via the food or water supply. The more global our food supply chain becomes, the less control we have over contamination. Also, the more aggressive our weather becomes, the more potentially contaminated our water supply can become both through drought, which concentrates pollutants, and adverse weather events, which can affect the quality of ground and well water. We have yet to understand the full impact of these events on ground water, local rivers, streams and our oceans. Know where your food comes from. Get your water tested, annually.

What Goes in Which Produce Drawer and Why; Or, How to Make Your Grocery Purchases Last

We have all fallen victim to purchasing beautiful produce (especially bunches of leafy greens and baskets of berries) only to return to the refrigerator a few days later to witness the soggy brown, moldy mess they've become. Here's why. As some produce ripens, it releases ethylene gas. The more ethylene released, the more likely that produce stored nearby in the refrigerator drawer will rot. Depending on the fruit or vegetable, different amounts of ethylene gas are emitted. To get the most for your money, be sure to buy the best produce you can afford and put it in the correct drawer—or on the counter, as indicated.

As a rule of thumb, the *high humidity drawer* has no controls and will help preserve moisture in your produce and prevent it from wilting. We put leafy greens and salad greens in this drawer, as well as citrus. The *low humidity drawer* often has a tab to partially vent it, enabling the destructive ethylene gas to escape. Most of your fruit belongs in this drawer, with the exception of fresh berries that you'll want to store in higher humidity.

What Goes in the High Humidity Drawer?

(Tabs closed all the way to preserve moisture)

CRUCIFEROUS VEGETABLES

Leafy greens: arugula, bok choy, collards, dandelion greens, kale, mizuna, mustard greens, radish greens, turnip greens, watercress

Hearty crucifers: Broccoli, Brussels sprouts, cabbage (red, green, Savoy, Napa), cauliflower, kohlrabi

Hearty roots: radishes, rutabaga, turnips

ALLIUMS

Chives, garlic, leeks, onions

OTHER VEGETABLES
Artichoke, asparagus

HERBS
Basil, cilantro, dill, lavender, lovage, oregano, parsley, rosemary, sage, thyme

BERRIES
Blackberries, raspberries, strawberries

OTHER
Citrus, fresh ginger (store unpeeled in a zip-top bag)
Broad beans (also known as fava beans or pigeon beans)

What Goes in the Low Humidity Drawer?
(Open tab to allow gas to escape)
Avocado, apples, grapes, mango, pomegranate, watermelon

What Goes on the Refrigerator Shelf?
Fresh mushrooms—in a porous paper bag, or better still in their original packaging

On the Counter:
Tomatoes, garlic, jicama, onions, shallots, pomegranate—away from direct heat sources—so not on top of the microwave or toaster oven

The Best Way to Store Herbs—and Asparagus
Fresh herbs are pure joy for our genes and a must in the Genomic Kitchen. Earlier we saw that traditional diets among our long-lived Sardinian and Ikarian friends include a wide variety of fresh herbs. We therefore use them in as many ways as possible in basic recipes and dishes. That being said, unless you are growing your own herbs, they can be expensive so you need to protect your precious investment. Here are some tips, the very first of which is: don't wash them until just before use.

Here we are principally talking about rosemary and thyme. As some herbs mature, however, their stems can become thicker, and you need to pick the leaves from them before using in cooking. If your herbs have thick stems, place them on a slightly damp paper towel and roll them up loosely into a cylinder. Put the cylinder in a plastic bag and store in the high humidity drawer.

DELICATE HERBS WITH SOFT STEMS AND ASPARAGUS

Snip off the ends of the stems and remove any wilted or bruised leaves. Store them upright in a glass jar with ½-1 inch of water (depending on stem length). Then cover the jar with a plastic bag and secure with a rubber band. Store on the refrigerator shelf.

BASIL

I buy and use as soon as I can. I leave the basil on the counter in a plastic bag if I am going to use it within a day or purchase. If not, use the same technique as delicate herbs, but keep the glass jar on the counter. Best rule of thumb is: use as soon as possible, or pluck as you need leaves from a plant pot on your deck or window ledge, if you are growing them.

What Goes on the Other Refrigerator Shelves?

NUTS AND SEEDS

All. (See notes in freezer section.)

All nuts and seeds should—at minimum—be kept in the refrigerator or the freezer, unless you plan on eating them right away. See my notes about them in the freezer section shortly. The only exceptions are nuts sold in their shells such as peanuts and walnuts (but not pistachios). These can sit out at room temperature although some experts suggest storing in a sealed container. My advice: refrigerate or freeze all nuts and seeds unless you intend to eat them within a couple of days. In other words, don't let them linger.

PERISHABLE ANIMAL AND SEAFOOD PRODUCTS

Beef, lamb, pork, poultry game, eggs (see sidebar)

Herring, mackerel, salmon, sardines

FERMENTED AND CULTURED ITEMS

Cheese, kefir, kombucha, yogurt

Miso, tempeh

OTHER:

Wheat germ

To Wash or Not to Wash Your Eggs? That is the Question!

Around the world, many cultures store their eggs on the countertop, while in the US, we refrigerate ours. Why? It comes down to production practices which can impact safety from facility to table. Also consumer acceptance. When hens lay eggs, a protective layer, called a bloom, is added to the outer shell. This layer can prevent or mitigate harmful bacteria penetrating the eggshell. But eggs are not always laid in squeaky clean environments, so in the US, if they are not farm-direct, meaning their origins are from mass-production facilities, they are washed. Washing creates pleasing aesthetics for the consumer and also rids the eggs of possible contaminants. The issue then is that the protective bloom-layer is destroyed, the outer shell is weakened, and the egg thus subject to penetration of harmful bacteria such as salmonella. This being the case, the eggs must now be kept refrigerated to prevent contamination. So, once again, know thy producer. If you are purchasing eggs from a local producer, ask him or her about their hens, the laying environment, and whether the eggs have been washed. After that, the choice is yours.

What Goes In the Freezer?

I can't tell you how many times I walk into houses and see beautiful nuts and seeds stored in the pantry at room temperature. Nuts and seeds are not only delicious nutrient powerhouses, but they are all prone to rancidity because they are high in fat. Rancid fat and your genes do not work well together. Rancid fat has a way of driving oxidative stress, the opposite of what we are trying to achieve with good food. Nuts and seeds are also not cheap. Protect your investment and put *all* your hulled nuts and seeds in the freezer, or at very minimum in the refrigerator. Be sure to store them in a glass container, or freezer-friendly bags to add extra protection. I have found that even bulk-dispensed nuts and seeds, which have less time in the store and are sold more rapidly, can also develop off-flavors and start to taste rancid if they are not adequately protected in the freezer. Protect them well! Exception: nuts that are still in their shells can be stored at room temperature if you reach for them frequently!

Why do hulled nuts and seeds go rancid? Because their protective antioxidants have been removed during processing. This renders them more susceptible to oxidation which leads to rancidity and those off-flavors. So do yourself a favor and put your nuts, seeds in the coolest darkest place—your freezer!

NUTS AND SEEDS

Brazil nuts, walnuts, flaxseed, hemp seed, sesame seeds, sunflower seeds

My (Long!) Journey to Understanding the Uses of the Toaster Oven

Growing up in England, our houses were small and the kitchens, too. We had the usual range and often a grill that sat above the range. This would be akin to a salamander in a commercial kitchen, with a gas or electric coiling heat source above a rack, allowing the chef to grill, or cook food at a higher temperature using direct heat overhead. In England, the grill would be used to cook bacon, sausages, or put a crust on a shepherd's pie, etc. But one thing we never had was a stand-alone toaster oven. We had toasters for making toast, but not a "toaster oven." It has taken more than 20 years of living in the US to understand the value of this thing. I had only seen it used to toast bagels, toast, or—heaven forbid—Pop-tarts.

When I moved recently, the kitchen came with a toaster oven hidden behind "secret doors" in the kitchen. Why, why, why would anyone put this in that space, I exclaimed? The Cuisinart or the blender, but not a toaster oven!! Then I learned that you can use it to toast nuts and seeds and save on heating the large oven for such a small task. You can bake bread in it, though the results are not always as successful as baking in the large oven, I must say. You can bake a cobbler in it, broil a dish, roast asparagus, or any vegetable. In short, I discovered that you can do most things you would otherwise do in a large oven, but save time and money with (pretty much) the same results. How was I to know? If you happen to have one, use it.

What Goes in the Pantry?

CANNED OR POWERFUL DRIED LEGUMES

Broad beans, chickpeas, lentils, soybeans (can also be frozen)

When I researched foods with a broad spectrum of vitamins and minerals, these four legumes were standouts. Put them on your pantry shelf and use them. No need to eschew other legumes, but be sure to have these either canned or dried. Their nutritional diversity and culinary versatility will serve you well over and over again.

CANNED OR PACKAGED SEAFOOD

Mackerel, salmon, sardines, tuna

You know seafood is always best fresh if you are close to the source and know its age. But for many of us, packaged seafood is a handy staple. It can also be an economical alternative to fresh. If you cannot buy fresh, go for quality and sustainably harvested brands versus the cheapies. Put the bucks out there and purchase canned seafood from a company such as Wild Planet or Vital Choice. Choose companies who are dedicated to sustainable fishing methods and environmental stewardship. And be sure to refer to resources that keep you abreast of heavy metals in seafood like the wallet card we learned about earlier.

DRIED STAPLES

Dried mushrooms, particularly shiitakes, porcini, and matsutake

These are a useful alternative to fresh mushrooms and last a lot longer too. Serious flavor and deep nutrition accompany any rehydrated mushroom—and the hydration liquid can be recycled into a delightfully flavorful stock used in the dish you are creating.

GREEN TEA

A potent source of the bioactive Epigallocatechin Gallate (EGCG). A wide body of research supports the relationship between this potent bioactive and its impact on genes and pathways that prevent tumor formation (as in some cancers). EGCG has also been implicated in lowering the efficiency of genes that are involved in lipid metabolism that may lead to atherosclerosis. Green tea is thus both beverage and cooking ingredient in the Genomic Kitchen.

NUTRITIONAL YEAST

This food supplement adds nutritional and flavorful diversity to many dishes. Used extensively in vegan and vegetarian cuisine, it confers a nutty or cheese-like flavor to dishes, without the dairy component. Rich in B vitamins, it provides critical nutrient support to the essential transmethylation and trans-sulfuration cycles in the body, which you'll learn about shortly.

SEA VEGETABLES

Like the legumes in the Genomic Kitchen, sea vegetables (such as dulse, kombu, nori, and wakame, to name a few) are powerful base ingredients with huge nutrient diversity. Many of you will be familiar with nori—the dried "wrapper" around your sushi roll. I am not suggesting you suddenly start making your own sushi rolls unless you want to (great fun). But a simple way to include the nourishment of sea vegetables is to include them in your salt shaker. Here's how. Choose a coarse, whole crystal sea salt, like coarse kosher, preferably one you can buy in bulk. Put a sheet or two of nori with a cup of salt into your blender then pulse away until nori is a fine powder. Pour the mixture into your salt shaker and bingo!, you have just created a nutrient-rich condiment that is naturally iodized, as well. Asian groceries often sell this salt pre-made. Bottom line: find a way to put sea vegetables on your pantry shelf.

SPICES

Turmeric. I cannot say enough about turmeric, among so many other spices. We focus on turmeric in the Genomic Kitchen because we have the most supportive evidence about the bioactive it contains, curcumin. As we saw in previous chapters, this biactive works powerfully with your genes, particularly in turning off pathways leading to inflammation. Turmeric can be used as a single spice, or as part of spice blends, mostly curries and masalas. (Whole turmeric root is now turning up in markets too.) Whichever way you choose to use turmeric, just buy it and use it, *a lot*. In fact, you should be buying a nice range of spices and use them. *All of them*. The more we learn about the power and potential of spices and how they interact with our genes, the larger our spice cabinet should grow. And don't shy away from spice blends either. Chinese Five Spice blend, *Ras el Hanout, Dukkah, Garam Marsala,* and even the familiar mainstay *Herbes de Provence* are beautiful ways to add both flavor and a food-gene punch to your pantry and your food.

Hot Chili Peppers: The active component in hot chili peppers is capsaicin. Beyond stimulating your tastebuds and producing heat euphoria in your meals, researchers have uncovered a number of mechanisms by which capsaicin improves your health by speaking to your genes. Most notably, capsaicin is linked to biochemical pathways that stimulate endothelial vasodilation, improving the flexibility (versus constriction or tightening up) of blood vessels. Think of capsaicin as a bioactive that improves the structure and function of your blood vessels. Capsaicin has also been linked to improvements in fat *thermogenesis,* the process that taps your fat reserves for energy. Some research has also suggested that capsaicin plays a role in appetite control, signaling to your appetite-control centers to shut down the "eat more" mechanism. These are emerging areas of research, all of which are linked to capsaicin through its cell signaling, or nutrigenomic, capability. So don't by-pass the cayenne hot chili flakes anymore.

GINGER:

Don't leave the store without it! Fresh or dried, this potent, spicy root or dried spice adds heightened flavor to so many dishes, yet its true power is invisible to your taste buds. The bioactive component of ginger is gingerol which sends a powerful "shutdown" signal to master gene Nf-kB, a gene you will be learning a lot more about in this book. By blocking Nf-kB, you (in essence) shut down a cascade of inflammatory responses initiated by the IL-1β, IL-6 and TNF-α genes. Other research has shown how ginger helps us combat oxidative stress, and may also play a role in mitigating cancer. While these are complex mechanisms that are beyond the scope of this book, know that the fundamentals of ginger's potency lies in its ability to turn off inflammatory responses, and turn on the processes by which the body mitigates, or rids itself of, oxidative stress. More about that later, as well. For now, don't be caught empty-handed.

Recipe Break: A Tasty, Protein-rich Breakfast "Cereal"

I love this grain-free, protein-rich breakfast "cereal" which I adapted from an original recipe that appeared in Brendan Brazier's book, *Thrive* (highly recommended). You can mix or match this any way you want. I guarantee this cereal will fuel you from breakfast to way past your normal lunchtime!

GRAB THESE:

½ banana, sliced

1 pear or apple, peeled or not, but cored and grated

1 date, pitted and chopped

¼ cup toasted almonds or pecans, roughly chopped

1 tablespoon ground flaxseed (or whole chia seeds or flaxmeal)

½ tablespoon peeled, grated fresh ginger (about 1-inch slice)

1 tablespoon raw cocoa powder or carob (optional)

DO THIS:

1. Easy: mix everything together.

2. Eat as is, or serve with yogurt, kefir, milk, or dairy alternative.

Another Word about Herbs and Spices

Many of you use diverse herbs and spices in your kitchen, many kept handily on a rack above or next to the stove. Some rarely used ones— chervil, mace, sage, anyone?—have been languishing for years. Here is the nitty gritty. Heat causes both herbs and spices to lose their flavor. Light causes them to lose their color. Time causes them to lose both. Cool dark places retain both flavor and color, although opened spices will lose their power eventually. What we assume is that the same loss that leaches flavor, color, and pungency also impacts bioactive potential. So for any herb or spice, better to keep it well and keep it fresh and use it often.

Keep your herbs and spices away from heat and light. Better still, replace what you have with fresher versions. Many health food stores and natural groceries have bulk dispensers where you can buy a few tablespoons at a time. Buy what you need for a week or two or for that single recipe you found online. Empty your old spice jars and fill them with fresh spices, then *use them*. If you have too much of any one spice, store it in the refrigerator or freezer and use it as needed. The bottom line is that spices are hugely important, just like herbs, in the Genomic Kitchen. Treat them like your olive oil and crucifers, as revered ingredients to be coveted and respected.

Liquid Staples

HONEY

This natural, nutrient-laden sweetener is also rich in many bioactives, including quercetin, chrysin, and luteolin. These are among a growing number of bioactives which appear to have the ability to shut down enzymes which alter gene expression in ways not conducive to health. A natural "blocker" is one way to look at what the bioactives in honey do. A growing body of research has also linked chrysin with an anti-cancer effect in cell cultures in the laboratory setting. Chrysin not only appears to prevent the growth of tumorogenic (tumor-forming) cells, but also induces the death of the same. Why not put it in your pantry for these reasons but also as an unprocessed sweetener?

OLIVE OIL

Olive oil, like fresh produce, is liquid gold for our genes. If I could choose one food that receives the crown for its nutrient and bioactive diversity, it would be olive oil. While it is not used in all cuisines around the world by any means, it is possibly the most researched fat in the world. We understand its nutrigenomic potential, therefore we use it daily and give it a central place in the pantry. Don't go for cheap. Go for the best. This is liquid medicine in your kitchen. Buy the best. Try different varieties. Go local if it is available. It will serve you well.

Olive Oil: One of the Best Foods for Your Genes and Your Health

If you search nutrient databases for food rich in bioactive polyphenols, unrefined olive oil is in a class of its own. Polyphenols are a large family of non-nutrient compounds called phytochemicals which have healthful properties. Names of these families include flavonoids, carotenoids, and stilbenes. *Bioactive* polyphenols frequently refer to the antioxidant capability of this family. (Remember, antioxidants stabilize destructive free radicals which cause havoc in your cells and tissues.)

The wide array of polyphenols in olive oil provides a rich antioxidant protective barrier for the fragile fats that make up the oil. Remove the protective polyphenols and you get rancid fat and a circus of free radicals. Research into heart disease (among others) suggests that oxidative stress, caused by free radicals, and inflammation are a root cause. Bioactives can fire-up genes which help quench the free radicals that cause oxidative stress. Uncontrolled oxidative stress (untamed free radicals) damages the cells and tissues of your arteries. Bioactives also interrupt signaling receptors on cell membranes which, if turned on, set in motion a cascade of inflammation damaging the lining of arteries. **Luteolin**, found in olives and olive oil, (and robustly in some Mediterranean herbs such as sage and thyme) is one of those good interfering compounds.

While we still have much to discover about the food-gene capabilities of many bioactives, rest-assured that foods replete with these components will be far better for your health and longevity than foods that contain none of them. Buy high quality olive oil—it should cost at least $40 for a 3-liter can. Put it in a cool dark place. Use it.

How to choose the best olive oil? Know thy producer! Now, most of us don't. If you can buy from a small vendor—great. If not, learn how to buy olive oil. If you want to know my favorite, it is Milestone Olive Oil from Greece. I know the producer. I know the olive. I know the land. Another wonderful oil is Olea Estates. This oil is produced in the Sparta region of Greece and sold in single bottles or larger 3-liter cans. Pruneti is another personal favorite of mine. Go forth and seek out the best olive oil you can buy. Your genes and health will thank you.

RED WINE

Perhaps not all of you will choose to add red wine to your table or to your cuisine. For those of us who do, we know the rich flavor this lends to cuisine and how it pleases the palate. Red wine, along with grapes (particularly red grapes), is the most powerful source of the bioactive resveratrol which upregulates (turns on) the Sirt-1 gene that we'll soon learn is a powerful player in blood sugar and fat metabolism. Choose your best wine or keep those grapes on hand.

VINEGARS

These add both flavor and acidity to cuisine and are crucial to the many vinaigrettes we'll be making as part of the Genomic Kitchen culinary collection of recipes and techniques. You can certainly make your own vinegars, but I say go to the store and grab some vinegars to add to your pantry. Staples include red wine vinegar, rice wine vinegar, sherry vinegar, balsamic vinegar and apple cider vinegar. Start with a good red wine vinegar for everyday use. Here's a trick: make your own balsamic vinegar reduction for fruit. Pour one cup of quality balsamic vinegar in a small saucepan and gently bring to a boil. Reduce the heat to a lower simmer and cook gently for about 20 minutes until the liquid has reduced by half. A taste at this point, and, *voilà,* you have an aromatic syrup that can be poured over fresh peaches, apples, pears, melon, and beyond. Or, for a deeper dive into delicious, melt a square of dark chocolate into the reduction and then pour over your fruit. (You may have to reheat leftovers to liquefy—if there are any leftovers.) I promise this will become a staple in your recipe repertoire!

What You Won't Find in the Genomic Kitchen

The Genomic Kitchen utilizes a specific group of ingredients that form its core. I chose them according to the principles I'm sharing in the book. They either contain bioactives which are central to creating a food-gene relationship, or they support the body's ability to absorb and use vital nutrients that support the work of genes and your essential biochemical pathways. This does not mean you must stock your kitchen exclusively with these ingredients. It means they are foundational and should be used regularly and intermixed with other favorites and those you have on hand. Let's say you like to use butter or coconut butter for fat. Fine, use it. I do, too. Thai cuisine without a can of coconut milk in the pantry, for instance, just wouldn't be Thai cuisine. My pantry is not complete without it either. I didn't mention a lot about sweeteners either. I use honey as a core sweetener, but often prefer the texture and fluidity of pure maple syrup in recipes. Occasionally I like palm sugar or coconut sugar as well.

This being said, there are some ingredients I do not include in the kitchen and, frankly, do not use. I don't have packaged cereals in my pantry because I make nut-based cereals or granola from scratch, controlling the ingredient inputs. There are no bottled salad dressings on the shelves or in the refrigerator. I again make them from scratch, choosing to focus on quality ingredients that also have nutrigenomic potential. I do have a variety of quality flours because I like to make bread from scratch and freeze it. I don't have cake mixes, batter mixes, or "helper" ingredient mixes. If I want to bake, I make from scratch. Of course, sometimes the demands of life and busy families make life a tad easier with these convenience products. While I don't advocate them, I do understand why some of you may use them. Just keep them as a last resort and focus on whole ingredients and simple meals made from scratch to support your best health.

Some Tools for Your Kitchen

Just as you don't need to dump everything in your refrigerator or pantry, the same goes for your existing equipment. Sure a good knife and a sturdy pot to cook in are essentials, but you don't need to rush out and replace everything. When you can, however, buying quality equipment will serve you over the long-term both in ease of use and years of use. For now, we are going to use the "Keep It Simple" strategy here.

Working in my kitchen, I notice that I use the same equipment over and over and sometimes need more than one of the same item. Here is a list of essentials, followed by some notes, of my frequently used items, which I recommend you have.

3 large cutting boards (I do use plastic cutting boards for ease of cleaning)

2 large stainless-steel mixing bowls

3 mid-sized stainless-steel bowls

5 small (1 cup size) stainless-steel bowls

Small glass or stainless-steel bowls (2 oz. size) to hold measured ingredients in small quantities, like spices

Assorted measuring cups: nested cups are fine, ¼ cup to 2 cup is great

1 microplane (get the long thin one that looks like a wood file)

Juice press

Large box grater

A chef's knife (7" or 10")

A paring knife (about 5")

Knife sharpener, or get them done professionally and regularly

Peeler

Gaggle of spatulas (for blending, stirring)

One pair each, short and long tongs for picking up or tossing food

Bench scraper (some refer to it as a pastry scraper)

2 cup mini food processor

8 cup food processor

Cooking pots: I frequently use 1-quart, 4-quart saucepan, and 6- and 8-quart saucepans. I also have a recently developed a love affair with all things Le Creuset, which seem to double as every form of cookware you ever need.

Skillets: I *love* All-Clad and Scanpan—but use what you have: 1 fry pan (low rim) and 1 sauté pan (higher rim)

A variety of serving bowls and platters, or double-up your mixing bowls as serving bowls

---

So that's the basic equipment list. Use what you have and add to your kitchen when you can. Anything else is supplemental. For instance, I love to use my pressure cooker to speed-up cooking and create one-pot meals. It is not an essential, but a "nice to have."

Bringing It All Together

As you read the remaining chapters, particularly the recipes and suggested cooking techniques, you'll come across many items from my ingredient staples and the equipment list. But here are some quick insights to get us started. To create food that connects to your genes, we use a combination of cooked and raw foods using a variety of equipment. Don't panic, this is not a raw food kitchen!

As you'll learn, we use a variety of techniques to connect ingredients to the food-gene relationship that make it deep and complex. For example, one easy way to do this is to make salads and slaws. To make them tasty, we chop or grate simple ingredients, then toss them with a dressing. So right there, you see the need for a cutting board, knife, whisk, grater, and bowls.

To make dressings tasty, we might use a microplane for zesting, or a citrus press for juice. We toss everything together using tongs.

To sauté greens, which we use a lot, you need your sauté pan.

To add the genomic connection back into your cooked greens, you'll need to chop one or two additional ingredients and add at the end of cooking. Once again, we're back to the knife, cutting board, and bowl.

Let's Make a Food-Gene Connection with Garden Greens

As you'll learn in the coming chapters, we like to use a mix of cooked and raw in our Genomic Kitchen to create the best food-gene connection.

Cabbage (or any other cruciferous vegetable) is a source of the bioactive sulforaphane. We access this bioactive by chopping or tearing up the cabbage and eating it raw. When we cook cabbage, or add an acid to it (in this case, lemon juice), it prevents the formation of sulforaphane. Now, there are ways to create a workaround in the kitchen to preserve sulforaphane, which we'll be talking about in the next chapter. For now, let's assume you are going to make this recipe and yes, destroy the precious sulforaphane bioactive by cooking the cabbage. What's a person to do? Well, in the case of this delicious recipe, if one part of the recipe prevents the production of our key sulforaphane bioactive (cooking the cabbage), we can add it in at the end of the recipe using another cruciferous ingredient—in this case, arugula or a handful of watercress. You then toss the cooked cabbage with these final ingredients or scatter a generous handful on top just before you serve, almost like a garnish. See, you can have your sulforaphane and eat it too! Try this for yourself with a recipe I slightly adapted from the great chef and cookbook writer, Deborah Madison, in her brilliant book, *Vegetable Literacy.*

Sautéed Lemony Red Cabbage with Fresh Herbs

GRAB THESE

3 tablespoons olive oil

1 medium (or 2 small) red onions, paper removed and cut crosswise into thin rings

2 small cloves of garlic, minced

5½ cups packed, very finely sliced red cabbage (1 1/4 lbs.)

About 1½ teaspoons sea salt

1 lemon zested and juiced (more if lemon is not juicy)

¾ cup mixed fresh herbs: dill, parsley, mint, basil, thyme

1 cup or more fresh watercress or arugula (pre-packaged is fine)

Black pepper to taste

DO THIS

1. In a large frying pan or wok (I prefer a wok) heat the oil gently over medium heat.

2. Add the onions, using tongs to turn and coat with oil. Cook for about 2 minutes to soften them.

3. Add the garlic, cabbage, and salt tossing them together with the onions. Cook lightly, using the tongs to move the ingredients constantly. You want the cabbage to be barely wilted or still toothy, but not raw.

4. Remove the pan from heat. Pour the lemon juice and zest over the cabbage mixture and toss to combine. At this point, taste and add more lemon juice as needed.

5. Now toss in the herbs and your extra handful of crucifers. Add some good grinds of pepper and you are ready to serve.

I love this recipe served warm from the pan, or as a cool leftover the next day. You can also add a dollop of yogurt-dill dressing, or even a yogurt dressing scented with harissa or cumin. A sprinkle of sesame seeds adds another dollop of nutrients (copper, manganese, calcium, phosphorous to name a few) if you want to power this up to another level! And yes, if you can't live without some heat, a few grinds of Aleppo pepper do the trick.

Connecting Your Kitchen to Your Genes

Now that you have insights into our core ingredients and our equipment essentials, it's time to move on to the "why" and "how" of the Genomic Kitchen. Over the next four chapters, I will introduce you to the set of principles we call MISE. These principles will explain the science behind the ingredients we are using, and more importantly, how they connect to your genes and your health. We organize the MISE principals into the Genomic Kitchen Ingredient Toolbox, which I will also slowly unpack for you in the following chapters. By the end of this book, you will have a clear understanding of why you are stocking your pantry and refrigerator with specific ingredients. More critically, through tips, techniques, and sample recipes, you will know how to walk into your own kitchen, grab the ingredients you need, and create a new relationship with your health just by cooking and eating. Welcome to a new food conversation. Now let's get started!

Chapter Five: Master Ingredients

Not all genes are created equal. Some, the ones I call Master Genes, are more influential than others because they have a deeper, more far-reaching impact on our health. Not only do they exert influence on the pathways and systems that control oxidative stress and inflammation (that you'll be learning more about in this chapter), they also direct a number of genes like the conductor of your biochemical orchestra. Because Master Ingredients are those that are closely related to the functioning of our Master Genes (see below), by eating them we keep them operating efficiently and, by doing so, care for our body at the most fundamental level.

The Science Behind Master Ingredients

Master Ingredients influence our bodies' Master Genes, those which deeply influence our pathways to health, and, subsequently, our longevity. These Master Ingredients work by reducing two factors in aging: oxidative stress and inflammation. Oxidative stress and inflammation are considered to be at the root of core metabolic imbalances from which most chronic diseases like heart disease, diabetes, osteoporosis, and cancer (for example) spring. There are several Master Genes, but the ones that most directly impact oxidative stress, inflammation, and metabolism are:

The Fire Hose, *or Nrf2*

Nrf2 helps put out the fire of inflammation caused by oxidative stress that occurs in the body. Now, for the pure scientists among us, Nrf2 is, actually, not really a gene, but something called a transcription factor that you can think of as a *gene helper*. It is encoded for by a gene NFL2E2. As you can see, this could get confusing really fast, and I don't want to lose you in the minutia! I'll tell you more about Nrf2 shortly.

The Fire Igniters, *or TNF-alpha and NfkappaB (Nf-kB for short)*

TNF-alpha and NF-κB can spark and further stoke the fires of inflammation. We can use food to stop that spark.

Master of Metabolic Universe, *or SIRT-1*

SIRT-1 helps us efficiently manage inflammation, control blood sugar, and, in the case of fats, affects how the body metabolizes stores, and uses them.

Oxidative Stress and Inflammation: Rusting Our Body

Late one night you are backing somewhat aggressively out of the driveway when you hit that inconveniently placed post that you have been complaining about for a while. You clamber out of the car (muttering a few choice words) to survey the damage and check to see if anyone witnessed your unfortunate maneuver. No one saw you. Your car, however, is dented, and the dent chipped the paint. If you don't fix the paint, eventually your car will start to rust. Why? Because the raw metal that the paint was protecting is now being exposed to oxygen. In a similar fashion, free radical oxygen— or the unstable oxygen molecules that are generated inside our bodies—actually "rusts" us, causing disease and premature aging. That's oxidative stress, which, when left unattended, can trigger inflammation.

Oxidative stress occurs when free radicals overwhelm the body's ability to regulate them. Free radicals are generated as a normal product of metabolism, and the body actually has a built-in management system for them. They are also generated when the body is under stress, when we get sick, or when a disease like diabetes is active in us. Free radicals are oxygen atoms with an unpaired electron, which makes them highly reactive—since electrons like to travel in pairs—and thus capable of damaging other larger molecules like...proteins and DNA! Free radicals will rob other molecules of their electrons to stabilize themselves, causing a domino-like, or cascading damaging effect on cell membranes and tissues. Free radicals are also generated by stressors such as toxic environmental contaminants, pollution, radiation, and tobacco smoke, to name a few. When the body is unable to sequester, extinguish, or otherwise counteract the effects of corrosive free radicals, tissue damage and inflammation set in.

So while the bad news is that your body produces free radicals all the time as a byproduct of normal metabolism, the good news is that it can use antioxidants from your diet as well as from its own internal system to neutralize them. Dietary antioxidants, like the anthocyanins in a thick-skinned wild blueberry, can stabilize these pesky free radicals, rendering them harmless to the body. But the body actually has a unique and far more effective system to accompany the work of dietary antioxidants. Let's take a look at that system.

Nrf2 to the Rescue

We want to be able to nutritionally support processes in our body that protect us from oxidative stress and inflammation. Remember in Chapter Two we learned how Nf-kB initiates the first step in gene transcription? It is often referred to as a gene, but technically following scientific nomenclature it is really a gene helper or transcription factor. Well, Nrf2, often referred to as a gene, is an equally powerful transcription factor. It helps us to manage oxidative stress, which can lead to inflammation if left uncontrolled. How do get Nrf2 to work for us? We can eat very specific ingredients containing bioactives that activate Nrf2. Upon activation, Nrf2 induces the expression of genes whose protein products, or more correctly enzymes, act as powerful antioxidants. In essence, when we eat ingredients that support the work of Nrf2, we produce antioxidants that make us a more efficient manager of oxidative stress and inflammation. And, Master Ingredients contain important bioactives that activate NrF2. Let's learn how.

Putting Out the Fires of Inflammation

PART 1: TURNING ON THE FIRE HOSE

If your house was on fire, would you want to put out that fire with a firehose or by throwing teacups of water at the flames? A firehose can deluge your house with thousands of gallons of water. A teacup would have absolutely no impact. Of course, your answer will be a firehose.

Your body is not on fire. Not directly, but it is trying to combat the many little embers of oxidative stress caused by unstable free radicals that arise just from the body doing its normal housekeeping. In order to Do This, the body needs antioxidants. Many of us have learned that certain foods (those containing vitamin A, C, or E, for example) are powerful sources of antioxidants. Yes, these

vitamins are important antioxidants, but they are akin to our teacup. They can extinguish free radicals in the body at a rate of one
to one: one vitamin A molecule to one free radical molecule. Not.
Very. Effective.

In contrast, we can turn on the body's firehose by choosing
Master Ingredients containing specific bioactives that activate
Nrf2. (Nrf2 bears the rather unwieldy name Nuclear factor
E2-related factor 2, hence the abbreviation.) You just learned
that Nrf2 activates genes to produce enzymes that douse the
fire from free radicals in your cells, akin to a firehose. Nrf2 also
influences the production of other enzymes instrumental to our
healthy processes (detoxification for example). Perhaps none are
more powerful than the range of enzymes that help us effectively
manage oxidative stress. The most of important of these enzymes
in this case are Superoxide dismutase (SOD), catalase (CAT), and
glutathione peroxidase (GPx). These three enzymes work together
in a biochemical circuit to extinguish the fires of inflammation at
a rate of millions of molecules per second. Compare this to our
dietary antioxidants that operate on a 1:1 (or teacup) basis! Sure,
the body utilizes both antioxidant systems, which work together
synergistically. That being said, by activating Nrf2, we turn on the
body's most powerful internal free-radical management system.
So why not do it!

Deep Dive into How Nrf2 Turns Firehose Genes On

Nrf2 is essentially dormant, or in the off position, until your body sends a signal to turn it on or activate it. Turning Nrf2 on frees it from its dormant, or bound, position in the cytoplasm of the cell, enabling it to move into the nucleus or heart of the cell. Think of the cytoplasm as a liquid ocean surrounding an island, the nucleus. In the cytoplasm, Nrf2 is bound to a "keeper," called the KEAP1 complex, which you see on the top left hand side of the graphic. Bioactives such as sulforaphane and quercetin, which we find in the Master Ingredients in the GK Ingredient Toolbox, can send that important signal to release Nrf2 from its KEAP1 gatekeeper. At the graphic's bottom, we see that, freed from its gatekeeper, NrF2 can now move into the nucleus and bind to another platform or complex called ARE, or Antioxidant Response Element. On the far right: once Nrf2 binds to the ARE it initiates gene transcription which results in the production of proteins listed in the purple box. Bingo, we turned a gene on! There are lots of interesting explanations of the power of NrF2 on YouTube. Just search for "NrF2 and oxidative stress," or have a quick listen in here.

Putting Out the Fires of Inflammation

PART 2: SUPPRESSING THE FIRE IGNITERS

You just learned how to turn on your body's firehose to extinguish fires that can be caused by free radicals. The question now is: can you also prevent fires from inflammation from occurring in the first place by blocking igniters that start fires? This is where things get exciting. You see, transcription factors don't always turn a gene on. They can also turn a gene off, or act like a dimmer switch, telling the gene to slow down and sometimes halt protein production.

Our major fire igniters are NF-κB and TNF-alpha. Remember that NF-κB is often referred to as a gene, but to be scientifically accurate, it is really a transcription factor. NFKB1 is its gene parent. Transcription factors act as gene helpers or "assists" nudging genes to get on with their business of protein manufacturing. While NF-κB (and Nrf2) are playing an assist role in our genetic conversation, you will be pleased to know that TNF-alpha is its own gene and does its own work without an assist!

It is important to know that NF-κB and TNF-alpha play a critical role in managing the body's immune response and can act as a first responder to cellular threats. Think of them providing a first line of defense against injury, such as when we cut ourselves or get into an accident. Where we get into trouble is when NF-κB and TNF-alpha are put into overdrive by less noxious, ongoing threats to the body, such as environmental burdens, pollution, toxins, life stress, and an overabundance of free radicals.

In overdrive or "Fire Igniter" mode, NF-κB and TNF-alpha can actually *initiate* inflammation in the body. Think of TNF-alpha as the ignition switch and NF-κB as the spark. Working together, the effect in the body is to start a blaze of inflammation through the production of significant numbers of inflammatory molecules, such as cytokines. Because cytokines are specialized cell-signaling molecules, they can amplify immune responses; thus increasing the movement of white cells towards sites of inflammation, whether caused by infection or trauma. Remember, inflammation is considered a root cause of disease. We do not want to start fires of inflammation, and we don't want to stoke smoldering embers either. Yet many of us are walking around with both burning inside of us.

So what is a person to do? The answer lies once again with our friendly bioactives that we consume in specific foods. The good news is that some of the *same bioactives* that turn on our important Nrf2 firehose to extinguish those free radicals can also block the TNF-alpha and Nf-kB switches which ignite inflammatory fires. How cool is that?

Master Ingredients contain the bioactives that helps us turn *on* Nrf2 and turn *off* TNF-alpha and Nf-kB, thereby helping us manage both oxidative stress and inflammation. And here's a little nugget that leads right into our next Master Gene, SIRT-1. It turns out that one of the many attributes of SIRT-1, is its ability to also turn off NF-κB. It's like a see-saw. You turn on one gene and it blocks another. Ah ha! The body is very wise and replete with these built-in internal biochemical checks and balances. You'll be reading about SIRT-1 very shortly.

Nrf2: Beyond the Firehose to Mass Evacuator

One of the things that firefighters do during a fire is evacuate people (and pets) to safety from a burning structure. The body has its own evacuation systems to remove toxic substances and return the body back to balance. Left to their own devices, toxic substances are akin to embers that get picked up by the wind and, unknown to us, kindle fires, often far from the source; in this case, throughout the body. The evacuation process here is the bodily process of ongoing biochemical detoxification. Rather than a diet fad you might have heard of, detoxification is a natural process. Yes, our bodies all have the mechanisms to enable the detoxification process. And despite what is posted in popular press by everyday folks and some clinicians too, some of us need a little more help with detox than others because we have SNPs in our detox pathways.

Nrf2 gives us a one huge leg up in the detoxification process. Here's how. As you recall, by turning on Nrf2, the body produces firehose antioxidants. But that's not all! Nrf2 also is instrumental in the production of enzymes that are involved in detoxification. These include the glutathione-S-transferase (GST) class of enzymes which help remove free radicals, environmental toxins, and heavy metals. Additionally Nrf2 can nudge another gene into action, namely NQo1. When activated, NQo1 encodes an enzyme with the unwieldy name, NAD(P)H dehydrogenase [quinone] 1, which we shorten to NADPH. The major role of NADPH is to detoxify or stabilize substances called hydroquinones. These are very reactive molecules that are extremely toxic to the cell and require stabilizing or neutralizing. Beyond detoxification, NADPH plays a vital role in protecting cell membranes, working together with antioxidants vitamin C, vitamin E, and co-enzyme Q10.

What I want you to see here is the connection between Nrf2 and the end processes of detoxification and the prevention of oxidative stress. The body has unique, yet complex biochemical management systems replete with checks and balances. We just need to enable and support those pathways and processes to help the body do its work. This we can accomplish through eating the right food.

A Serving of Inflammation with a Dollop of Cytokines

Your body engages the inflammatory process to repair cells and tissues and to remove damaged or dead cells. Inflammation can be acute or chronic. Acute inflammation is characterized by heat, pain, or swelling resulting from a cut, fall, or traumatic injury. Chronic inflammation is at the heart of disease and reflects an immune system constantly set in the "on" or "on-guard" position as a result of metabolic imbalance. Chronic inflammation may take the form of aches, pains, and general malaise, but not necessarily acute pain or swelling.

The inflammatory process is complex and involves many systems in the body. One part involves immune cells which release a variety of different chemical or hormone-like substances. For example, when tissues are injured by bacteria, toxins, or trauma, histamine and/or prostaglandins

(PGE1 and PGE2 are two specific types) are released, causing aches or pain. Blocking their production—which is what Tylenol and other common pain medications do—reduces the pain.

Cytokines are a different part of the inflammatory response system. They are small protein molecules secreted by cells that signal to other cells in the immune system. Cytokines are activated by macrophages which play the role of master vacuum cleaner in the immune system. Curiously, cytokines can be both anti-inflammatory and pro-inflammatory. Anti-inflammatory cytokines can actually repress the action of their pro-inflammatory counterparts. Pro-inflammatory cytokines unleash cascades of inflammatory activity. Think about it. It is a lot more difficult to put out fires igniting and raging in a number of different locations than it is to contain a single fire. Hence, turn off the system and treat the root cause rather than its result, the symptom.

SIRT-1: Longer Living Through Better Blood Sugar Control, Fat Metabolism, and Inflammation Regulation

SIRT-1 originally got a lot of press from research linking it to longevity. That research has fizzled out somewhat, being replaced with a much larger body of research shining a light on how this gene is involved in regulating oxidative stress, inflammation, and the metabolism of glucose (blood sugar) and lipids (fats). Think about that. We know that oxidative stress and inflammation are significant underpinnings of chronic disease along with genetic disorders such as Type 1 diabetes. We know that abnormalities in blood sugar and lipids are also hallmarks of chronic diseases present frequently in cardiovascular issues, Type II diabetes, obesity, etc. So if SIRT-1 is integral to processes that help us manage some of the above, then, yes, you can make a case for it helping us live a longer life. The devil is in the (evidence-based) details! While we will take a look at this Master Gene in some detail, those wanting a deeper dive can find a more complete discussion in Appendix B at the end of this book.

SIRT-1: Leader of Your Metabolic Orchestra

Two related mechanisms of gene activation are *acetylation,* by which genes are turned on, and *deacetylation,* which turns them off. Think of this as being able to flip a master switch on and off. SIRT-1 uses this master switch–process to turn on and off genes that not only impact inflammation, but also how the body regulates blood sugar and lipids. SIRT-1 is important!

Though I assign SIRT-1 to the category of Master Gene, it is important to be scientifically accurate and note that this gene does not work in isolation. It works closely in the body with other molecules (NAD+) and genes (AMPK) to direct its business. SIRT-1 is like an orchestra conductor and to conduct an orchestra we need a baton. NAD+ and AMPK are akin to handing the orchestra conductor the baton. Baton in hand, the performance can start. The baton, in other words, is an essential part of how the music unfolds.

And so it is with SIRT-1. Suffice to say that the science of how SIRT-1 conducts so many processes is complex, so I'll keep my explanations simple and provide you with a balcony view. As mentioned, if you want to take a deeper dive into the science describing the mechanisms of SIRT-1, you'll find them in the SIRT-1 Appendix at the end of the book. If you are further intrigued by this fascinating gene, then I recommend you follow the work of Dr. Satchidananda Panda at the Salk Institute for Biological Studies and Dr. Valter Longo at the University of Southern California Longevity Institute.

Finally, just a note that, while it may seem that SIRT-1 is in many ways the "master" Master Gene, it is *one* of many genes that are integral to how the body manages metabolic processes related to blood sugar and lipid management. SIRT-1 is perhaps unique because it is integral to many processes that affect these aspects of our metabolism. Suffice to say that many genes are involved in the body's complex metabolic processes, so SIRT-1 is far from the sole player.

SIRT-1 as Blood Sugar Manager and Inflammation Gate Keeper

SIRT-1 is integral to how the body manages blood sugar. This Master Gene impacts the process in many ways, and in many tissues. Let me explain how, keeping this as simple as possible. Let's start with the pancreas. This organ is at the epicenter of how the body regulates blood sugar. Its beta cells produce insulin, the body's chief regulatory blood sugar and fat storage hormone. Uniquely, SIRT-1 protects the beta cells by suppressing signals that activate the pro-inflammatory transcription factor, NF-κB. Yes, you just read that and you already know about NF-κB. By protecting beta cells from oxidative stress and pro-inflammatory assaults, the production of insulin for blood sugar regulation is better guarded.

Outside of the pancreas and in skeletal muscle, liver and adipose tissue, research indicates an instrumental role of SIRT-1 in sensitizing these tissues to insulin. Insulin is a hormone that is instrumental in blood sugar regulation. In some individuals and health conditions such as metabolic syndrome, cells become resistant to insulin resulting in higher circulating levels of blood sugar. As blood sugars rise, the body produces more insulin leading to hyperinsulinemia. SIRT-1 appears to short circuit this continuous loop by stimulating insulin signaling pathways. You could say that the gene is essentially prodding insulin to do its blood sugar regulation job. Simplistic indeed, but you get the point.

Finally, back to the topic of inflammation. The hyperinsulinemia I mention is often accompanied by low-grade inflammation in the tissues of people with Type II diabetes—and likely in those on their way to developing this health issue. We can measure inflammation by looking at levels in the blood of pro-inflammatory cytokines such as interleukin-6 (IL-6) as well as inflammatory marker C-reactive protein (CRP) for example. You may have seen some of these terms on lab reports your doctor has ordered. Much of this inflammation is generated by adipose or fatty tissue which we could assign the name "Inflammation Central." In other words, excess fat (which we might otherwise think of as carrying around excess pounds) generates lots of pro-inflammatory molecules that hinder insulin in doing its job. By blocking NF-κB activation which produces inflammatory molecules like cytokines, SIRT-1 can again support better blood sugar management. Through science, we know just how important SIRT-1 can be. Studies show that when SIRT-1 is suppressed or deactivated, it correlates with insulin resistance and metabolic syndrome which go hand in hand with heart disease, diabetes, and chronic disease. (Metabolic syndrome is a very specific diagnosis defined by the Mayo Clinic as "a cluster of conditions—increased blood pressure, high blood sugar, excess body fat around the waist, and abnormal cholesterol or triglyceride levels—that occur together, increasing your risk of heart disease, stroke and diabetes.") Now you can understand just how important SIRT-1 is.

SIRT-1 as Fat Manager

SIRT-1 plays a unique role in its interaction with fat. Many of us have struggled in life to lose a few of those excess pounds. We'd love to be able to steer our body to grab that fat and burn it off. Enter SIRT-1 and its fascinating role. When SIRT-1 is activated, we can suppress another gene, PPAR-gamma (PPAR- γ) that actually shuttles fat into storage. Turn off PPAR- γ (thanks to SIRT-1), and your body can better access your fat stores and preferentially use them for energy, instead of depleting all your carbohydrate stores, or worse, start breaking down muscle and other protein sources for glucose. In other words, SIRT-1 can reverse fat storage and promote fat burning. Research also shows that when you reduce your fat stores, you also improve insulin sensitivity and gain better control over blood sugars and subsequently your health. Wow!

Your Body Weight, the Body's Clock— and the CLOCK Gene

Did you know that the body has its own clock? It's called the circadian clock. And we all have a CLOCK gene, too. No, I am not making this up! Our circadian clock, technically the Circadian Locomotor Output Cycles Kaput gene, essentially manages the storage and use of energy by the body in conjunction with changes in demand during the day and night. Researchers have shown that people who have SNPs on the CLOCK gene, or whose work requires them to work evening and night shifts (the usual dark sleep cycle), show changes in the natural metabolic cycles governed by the CLOCK gene. These are changes in the natural circadian cycle by which the body anticipates feeding and managing energy, and such changes disturb the usual rhythm by which the body anticipates and manages food and energy. Disturbances to this natural rhythm often manifest themselves down the road as weight gain, metabolic syndrome, and beyond. High-fat diets have also been associated with changes in how the CLOCK gene behaves, with the same deleterious outcomes.

It turns out that SIRT-1 is deeply involved in the body's circadian clock, acting like a "central pacemaker" as one research paper described it. In other words, SIRT-1 appears to play a role in supporting the body's natural circadian rhythm and keeping its metabolic rhythm playing in time with the rest of the biochemical orchestra. For many people working third shift, routinely pulling all-nighters, or living with a sluggish metabolism, SIRT-1 is really good gene to activate and keep well-regulated.

How to Get Me Some of that SIRT-1!

You may have heard of the phrase, *"hara hachi bu."* It was popularized by the Blue Zones research, but has been integral to Japanese life for centuries. The rough translation is, "Eat until you are 80 percent full." For you and me, this means, don't over-serve yourself and do stop eating long before you feel stuffed. This is not only logical for maintaining a healthy weight, but also for healthy genes. SIRT-1 does not work entirely alone, but as we learned earlier in conjunction with another gene, AMPK, and a molecule NAD+. When any of these genes detects a decrease in energy in your cells (meaning that we need more energy), a signal is initiated, the end result of which is activation of SIRT-1. In other words, when the body needs energy, SIRT-1 starts tapping your energy stores, namely glucose and fatty acids. It's looking to burn calories.

Here's the nitty gritty—and why I mentioned *"hara hachi bu."* Rule number 1: don't overeat! A hyper-caloric environment, oth-erwise known as "stuffing yourself" turns SIRT-1 *off*. If you turn off SIRT-1, you cannot benefit from its power to help you manage your blood sugars and your fat. *Don't overeat. Period.*

The second way you activate SIRT-1, and its symbiotic compan-ions, is by eating foods containing resveratrol. The best sources of these foods are red grapes and the wine or juice produced from them, as well as boiled peanuts. Now when it comes to wine, make a note for yourself here. The people of the Mediterranean are known for their longer lives. They do drink small amounts of wine, often daily. They do eat grapes. They don't drink grape juice. Americans drink more grape juice than wine and that juice is often sweetened. It is, in other words, a far cry from unsweetened red wine, drunk in very small amounts. Interestingly, the amount of resveratrol in wine (and peanuts for that matter) can vary, depending on soil conditions and—importantly—plant stressors such as weather, pests, or fungal attacks. Resveratrol is an antioxidant and a defense mechanism for plants. The more stressed the plant, the higher the level of resveratrol.

Here's another thing to consider: resveratrol has a number of different forms we call isomers. The most common isomers are cis and trans, terms you may be a little familiar with from our earlier discussion of trans fat and lycopene and cooked tomatoes from Chapter Three. Research shows that the dominant form in red wine is the trans form of the molecule, and it is that form that initiates gene activity. We have been focusing on the role of resveratrol in activating SIRT-1, however, other research indicates an anti-proliferation (read as anti-growth) in cancer cells role for the trans form of resveratrol. Specific to SIRT-1 however, resveratrol concentrations in other foods are not necessarily in the trans form, and therefore cannot activate SIRT-1. Red wine and grapes are therefore your SIRT-1 go-to foods.

And on that note, I have surely regaled you with enough science. I have convinced you that we are now in a new and exciting era of understanding how food works in the body. You—hopefully—agree with me that food matters. So let's go into the kitchen, and meet the food that works with Nrf2, Nf-kB, TNF-alpha, and SIRT-1 and figure out how to get it on your plate.

On the following page are the Master Ingredients we're going to be working with. Take a quick peek and then let's get into the kitchen and learn how to work with them.

Fruit

Apples
Blackberries
Blueberries
Elderberries
Grapes
Pink Grapefruit
Pomegranate
Raspberries
Strawberries
(Cooked)
  Tomatoes
Watermelon

Vegetables

Crucifers:
Arugula
Bok Choy
Broccoli
Brussel Sprouts
Cabbage
Cauliflower
Collard Greens
Dandelion Greens
Kale
Kohlrabi
Mizuna
Radishes

Radish Leaves
Rutabaga
Tatsoi
Turnips
Turnip Greens
Wasabi
Watercress
Other:
Carrots
Fennel
Alliums:
Garlic
Leeks
Onions
Shallots

Herbs

Basil
Chives
Cilantro
Dill
Lavender
Lovage
Oregano
Parsley
Rosemary
Sage
Thyme

Legumes

Broad Beans
  (Fava/Pigeon
  Peas)
Boiled Peanuts

Spices

Ginger
Hot Chili Peppers
Turmeric

Other

Capers
Green Tea
Honey
Olive Oil
Red Wine

Chapter Six: In the Kitchen with Master Ingredients

Welcome to the Genomic Kitchen. Glad you finally made it here. If you already peeked at the table of contents at the beginning of the book, you saw that I divided the elements of MISE into chapters explaining the science, and then chapters that take us into the kitchen for each component of our Genomic Kitchen Ingredient Toolbox. This should allow you to digest (pardon the pun) the science and then walk it into the kitchen. My hope is that the science and its practice are like edible chunks. For those of you dying to get hands-on in the kitchen, you can jump right in there and see for yourself what I am talking about. So on the kitchen side of things, I want to give you a little warning ahead of time.

You have probably gathered by now that this is not a cookbook in the traditional sense. In fact, I did not want to write a cookbook. My approach, gleaned from twenty-plus years working with food and nutrition, is that it is more empowering to provide you with cooking *strategies* to use with ingredients than it is to provide you with 100 or more recipes. In other words, if I teach you the "why" behind the ingredients in the Genomic Kitchen Ingredient

Toolbox—and how to prepare them in the kitchen—you can choose and adapt any recipe you want, including the ones you already prepare and love. As you will see, first I teach you the "why" in the science chapters that accompany each MISE category, and then we go into the kitchen with a companion chapter. In the kitchen, I teach you different strategies for getting these key ingredients on your plate so that their nutrients and bioactives are readily available for your body to use. Of course, I have included some recipes for you, but consider them as examples of how you can easily prepare different ingredients from the GK Ingredient Toolbox and not as the *only* recipes you should make. All the recipes I have included have been prepared and tested by hundreds of people in my culinary workshops across America for many years. They serve as ideas for you as well as a jumping-off point for exploring your own delicious ways to prepare meals that influence your gene expression.

Before We Get Started

Before I get into using the Master Ingredients, let me address a couple of things. The first is a question I get asked frequently: how many (or how much) of these ingredients do I need to eat so that they meaningfully affect my genes? This is a good question and one that many clinicians ask me. Why? Because their clients are asking them. And the answer is: it depends. Nutrigenomics is a complex field of science. What differentiates it from the nutrition science is that nutrigenomics operates on the philosophy that every person is unique and different. This means that the amount of a bioactive or nutrient that will affect your personal biochemical system is different from that of your kids' or your neighbor's. Most traditional nutrition recommendations before nutrigenomic thinking are based on population-based studies and reflect observations and patterns that apply to lots of people, not specifically to you. This does not make these recommendations wrong, not at all. But nutrigenomics is more specific and applies on a case-by-case basis. The bottom line here is that every one of us has different nutrient requirements based on our unique metabolism, exercise demands, and, of course, our genome.

All that being said, let me reply to the *first* part of that question about *how much should I consume,* in this case as it relates specifically to bioactives. Here's the rule: It's not about the quantity, it's about their constant presence in your diet and thus their availability to do their job. Here's why. Bioactives in large quantities can be toxic. Surprise! They exert their effect by something called the hormetic response. This means that a *tiny amount* induces the reaction that we want. Translating this to guidance for you: eat foods rich in bioactives (from the Genomic Kitchen Ingredient Toolbox). By eating a variety of foods containing bioactives, you ensure both variety and a good supply of them. This will help you develop and deepen a solid relationship between your diet and your genes. Remember not all bioactives interact with the same genes. They are a bit choosey, just like you. So variety ensures their presence. The number of cups and calories you eat—this is not as relevant. Just grab the right ingredients and eat them, too.

The second thing I want to address pertains to the ingredients I include in the Genomic Kitchen. The toolbox of ingredients I present in this book focuses on foods containing bioactives and foods containing a robust variety of nutrients. Over its pages, I explain why you need to include as many of these ingredients as possible in your diet. Here's what I want you to know: the Genomic Kitchen has a *specific* ingredient focus, but not an *exclusive* ingredient focus. What I mean by this is we are ingredient *inclusive*, not ingredient *exclusive*. Our Genomic Kitchen Ingredient Toolbox is here to guide informed decisions around food rather than to have you exclude things you love.

The List of Master Ingredients

A quick glance at the Master Ingredients food list below shows you that you have lots of ingredients from which to choose. And there are many more to be unveiled in the following chapters. What you need to understand is why they are on the list, hence this chapter. Then you need to know how to prepare them. That's coming next. So take a quick glance again at the list and then we'll get to the good stuff of what to do with some of them.

Navigating the Master Ingredients

Probably the first thing you'll notice is that vegetables make up the longest list. Sorry about that! *Yes*, vegetables are important. Now take a closer look and you'll see that they are divided into Crucifers, Alliums, and Other. Crucifers (broccoli, cauliflower, kale family) provide the best precursor (or starting compound) for a powerful bioactive, sulforaphane. As we discussed in Chapter Five, sulforaphane is one of the most powerful activators of Nrf2 (the gene transcriptor that acts like a fire hose) that helps extinguish free radicals in cells. If you want to put out a fire in your system, you need to have crucifers on your plate. Bottom line. Now some of you are already protesting about the broccoli and Brussels sprouts, backing yourself against the wall and refusing to acknowledge their existence.

And rightfully so for certain individuals. If you have a variant on a gene family called TAS2R, you *might* be highly sensitive to the earthy or sulfur taste of these vegetables. You detect a "bitterness," as some clients tell me, that other people don't. You make a face when you put cabbage in your mouth, and others around you are asking what's wrong. It is true. It *is* in *your* genes. It explains why you love kale and your kids spit it out.

There are a couple of ways around this challenge. First, cooking crucifers by using oils and spices, or even roasting, can accentuate other flavors and mask the bitterness or earthiness that you pick up. Second, some crucifers are more pungent than others, and many are less pungent when young versus very mature. If you are a person who runs at the sight of Brussels sprouts, choose less pungent crucifers such as radishes, bok choy, or mizuna. Spicy crucifers include arugula and watercress, even radishes and turnips and their greens, depending on the season. Third, (and this is fail-proof as I have tested it with hundreds of people across the US) pair fruit, fresh or dried, with crucifers in recipes. Fruit adds sweetness, but also distracts you from the pungency of crucifers by taking your taste buds in a different direction. If you want to see and taste this in action, make the Root Vegetable Slaw. You'll see how I pair an earthy cruciferous root with sweet-tart apples to lift the flavor and then pull all the flavors together with an orange citrus vinaigrette. Arugula and ripe peaches or strawberries with just a hint of vinaigrette is also a favorite. If you are a beetroot avoider, check out what I have to say about them and fruit when you get to Chapter Eight, Influencers in the Kitchen.

Finally, if all else fails, you *know* you need your crucifers because they are essential to your food-gene relationship, so throw them into a fruit-based smoothie. You can grab a handful of leaves fresh or from a pre-packaged sauté or salad blend. Just add them to your favorite smoothie. Or make a pesto with kale or other leafy crucifers whose flavors are disguised by the stronger herbs, oils, and nuts that you would normally use in the recipe.

The second category of vegetables is the allium family—shallots, onions, scallions, and, yes, garlic. I have found that most people don't have a problem with these vegetables, although some folks cannot tolerate some of them. That's okay. I have an "out" for you. The allium family is a potent source of a bioactive quercetin. This powerful compound not only activates Nrf2 but it can act as a shield *against* activating the pro-inflammatory TNF-alpha and NF-κB duo. It's as if you get two for the price of one! Much American cuisine includes garlic and onions in some form, although usually cooked which, depending on the cooking method, can have an impact on the effectiveness of quercetin. We'll talk more about cooking methods in Chapter Nine so that you get the most out of your food-gene relationship. All that being said, if the alliums are not for you, you've still got choices. Radishes and their leaves, capers, elderberries, fennel and all of those herbs are in our Genomic Kitchen Ingredient Toolbox and are also good quercetin sources. And that is precisely why they are on the list!

Suffice to say that Master Ingredients include the best sources of the bioactives you need to interact with genes and transcription factors that help you manage free radicals, inflammation, and the metabolism of carbohydrates and fats. Now you know why we include these ingredients in our Genomic Kitchen Ingredient Toolbox. Try to incorporate some of them into each meal, or at minimum at least once a day. Finally curcumin, a core component of turmeric (but not of the spice cumin, despite the name)—use it everywhere you can, although don't drive your friends and family nuts. Like sulforaphane, curcumin has been robustly studied, but more specifically it is both an NrF2 activator *and* a potent antagonist for NF-κB and TNF-alpha. It is therefore excellent fortification information for your genes. Use it.

If you're wondering about dried turmeric versus the live root, I simply want you to *use* turmeric any way you choose. But if you like the details, it turns out that the root is actually more potent. Researchers found that roasting the root produced some additional healthy compounds as by-products. Analyzing these compounds, namely vanillin, ferulic acid, and 4-vinyl guaiacol, researchers discovered their role in turning off Nrf2, blocking the inflammatory IL-6 cytokine, and turning on an important gene, paraoxonase 1, or PON-1. PON-1 is significant because it detoxifies organophosphorus compounds, which you might be familiar with by their use in pesticides. How can you use the root in cooking? Certainly peel and grate it and throw it into smoothies or juices, which would keep it in its raw form. To use it in cooking, add grated root to stir fries or when you roast vegetables, for example. Some people even throw it into pumpkin pie or muffins. These are just ideas. Let your imagination take you where you want to go!

Connecting Your Genes with Your Plate

Let's start with some simple guidance before we get cooking to make sure we get those Master Ingredients onto your plate. Culinary Genomics is all about how to prepare ingredients to unlock their greatest potential. Whenever we work with Master Ingredients, we need to release and/or preserve the bioactives they contain. Bioactives in some of our ingredients are dormant and have to be awakened through chopping, tossing, and mixing them. Other cooking techniques actually destroy bioactives, reducing the ingredients' overall effectiveness. This is particularly true for our Master Ingredients. Let's look at some basics for you and then how to translate them into action.

Follow the Cooked/Raw Rule

Most bioactives are heat sensitive, which means that cooking can decrease their availability and activity. But as with everything, there are exceptions. Quercetin (think onions and garlic) and luteolin will withstand gentle cooking heat. Others, like some of the Nrf2-activating bioactives in turmeric are actually *enhanced* in the cooking process. You get the best nutrigenomic effect from your tomatoes when you cook them. Same with peanuts. Choose the *boiled* ones for greatest impact! Who knew? And to twist this a little further, the enzyme that helps us produce the potent Nrf2 activator, sulforaphane, is actually denatured or destroyed by heat! So what's a person to do? In Chapter Nine of the book, I include a chart to help you navigate these idiosyncrasies; but for now, approach your food with a cooked/raw philosophy. If you want some of your ingredients cooked, then cook them, but throughout the day, think of how you can balance cooked and raw to maximize food-gene potential.

Hack and Hold Your Crucifers

Sulforaphane is a powerful Nrf2-activating bioactive derived from the glucoraphanin compound that naturally resides in cruciferous vegetables. Note that I said "derived." Sulforaphane needs to be *created* or manufactured as it does not naturally exist in these veggies in their raw state. We produce sulforaphane by cutting (think thinly slicing greens or cutting broccoli into florets) or chewing crucifers to release an enzyme called myrosinase. Myrosinase converts glucoraphanin to sulforaphane. Here's the twist and then the rule. Myrosinase needs time and no heat to produce sulforaphane. If you immediately chop your cabbage or kale and start cooking it right away, not only do you not give myrosinase enough time to do its sulforaphane production work, but you destroy it, too. End result: little to no sulforaphane.

Here's how you work around this. Employ a "hack-and-hold" approach. If you cut up your crucifers and let them sit for 60 minutes, you give myrosinase enough time to produce sulforaphane. Unlike myrosinase, sulforaphane can withstand gently cooking techniques like steaming, quick stir-fries, and yes, even microwaving, not that I advocate microwaving all your veggies. The point here is that if you chop, rest, and then eat or lightly cook your crucifers, you will create the sulforaphane you need and get to eat it too! More on creating work-arounds in Chapter Nine.

Precut Versus Whole?

What about those pre-packaged salad or pre-cut raw vegetables mixes you buy at the store? Many of these mixes contain crucifers, so use them. Remember there is a law of diminishing returns. Yes, these mixes may contain sulforaphane (and many other excellent nutrients) after they are processed and packaged, but the potency of sulforaphane diminishes over time. If you want to use pre-packaged vegetable mixes, find the latest "Sell By" date so you pick the freshest and maybe secure some of that sulforaphane. Then, think about how throwing in a handful of fresh crucifers to refresh the sulforaphane content. I'll talk more about ways to Do This in Chapter Nine. For now, think about cutting up a few radishes, or adding a handful of arugula or watercress to that salad mix you bought. Or finely slice a little bit of fresh cabbage and add it to a pre-cut slaw mix. Remember, food is messaging your genes. You want to provide your body with the healthiest messages you can get.

Canned Produce or the Frozen Food Aisle?

Let's be clear: a mixture of fresh, canned, or frozen produce stretches the food dollar and meets the convenience factor. And in truth, when produce is picked and frozen or canned within a short time of harvest the nutrient content can mirror or even exceed fresh. How come? Because we have all seen wilted, flabby produce languishing in the grocery store; a sure sign that the nutrient-best of fresh has retired from the plant forever. Depending on the time from harvest to store shelf, nutrients lose their potency; therefore frozen or canned can be a more nutrient-dense option. That being said, when we're talking bioactives, high heat blanching or processing in production facilities can deplete them, as well as any water-soluble vitamins. Processing also changes the texture of vegetables. You'll never get the crunch or bite you will with fresh vegetables. So my advice is to use frozen and canned in smoothies, dips, or as an addition to slow-cooker meals or casseroles, and put your money into fresh for those salads and sides where you can access most bioactives at their potent best.

Use Acids Carefully and Prudently

Cooking acids can denature enzymes and render bioactives useless, just like heat. That means that by adding vinegar or citrus to our Master Ingredients we are diminishing the potency of their bioactives and weakening their efficacy. You can use acids such as vinegars and citrus zest and juice, but toss just before eating. For example, toss your salads at the last minute, or serve the dressing on the side to prevent acid from coming into contact with your ingredients until the very last moment. This is actually sound culinary practice, too, as oils in the dressing can literally drown your ingredients, making them limp and unappetizing.

So now you possess a good handful of the basic rules of working with some of your Master Ingredients, here's a quick list of how to get them on your plate so they work in tandem with your Master Genes.

Thirteen Ways to Help Crucifers Rock Your Genes—and Your Food

- Grab pre-packaged salads containing arugula, mizuna, baby kale, or all of them. Check for freshness. Use them abundantly.

- Eat raw pre-cut crucifers, party trays, or pre-packaged, with a dip. Check the preparation date for maximum freshness.

- Throw raw leafy greens (pre-packaged ones or from the farmers market is fine) on your plate as a bed for anything. Arugula and watercress mixes are excellent here.

- Make thinly sliced boy choy—leaves and stems—the base of your salad instead of lettuce.

- Throw thinly sliced leafy greens in every stew, risotto, slaw, or salad at the last minute.

- Eating pasta? Just before serving, throw in two handfuls of thinly sliced greens. The heat from the pasta will wilt them just right.

- Grain salads: you know, stir in thinly sliced greens!

- Smoothies: this is easy! Add a handful of cruciferous greens to every smoothie you make.

- Pesto: add a cup of cruciferous greens to your pesto base.

- Arugula: throw it on top of your pizza after it comes out of the oven.

- Process greens or cauliflower into dips.

- Make raw radishes finger-food and a part of your salads.

- Secret from Japan: Wasabi is a crucifer. Make the most of it when you eat sushi.

Three Template Recipes to Start You on Your Own Path

First, let's remember my broader guidance. Master Ingredients are part of your kitchen, but not your whole kitchen. When I have taught chefs and clinicians about bioactives, some become obsessed about preserving the bioactives and start worrying about cooking. Everything in moderation! Culinary Genomics is a *strategy* for eating over a lifetime. You want to prepare food that is both cooked and raw. Cooked food provides valuable nutrients above and beyond bioactives. Raw food will almost always deliver bioactives and nutrients. So balance cooked with raw. Let's take a look at three recipes to show you how this works. We'll get started with a basic salad.

Kale, Avocado, and Tart Cherry Salad

How It Works—Lemon juice is an acid, which helps break down the cell structure of the kale, softening it the same way sautéing will. Though acid and heat denature kale's myrosinase, meaning that it produces no sulforaphane, we need that acid or heat to make the kale more palatable. (An inedible dish won't do anyone any good.) In this recipe, we use lemon juice to soften the kale, but we add in daikon radish (a crucifer), right at the end to ensure you get your kale and your sulforaphane too! And for you kale-shirkers, give this recipe a whirl. I have had many kale converts after eating this recipe.

PREP: 20 MINUTES (PLUS 1 HOUR FOR KALE TO MACERATE).
SERVES 4.

GRAB THESE

1 bunch kale: easiest to use is Lacinato, washed

Juice and zest of 1 lemon (2 if the bunch of kale is big or lemon is not juicy)

¼ cup olive oil

½ teaspoon sea salt

¼ cup grated or peeled carrot, sweet potato or golden beet

¼ cup (or more) grated or peeled daikon radish

1 avocado, cubed

¼ cup tart cherries

¼ cup roasted sunflower seeds

Drizzle of honey as needed

DO THIS

1. Strip the kale leaves from the stem. Discard the stems. Tear or cut the kale leaves into bite-sized pieces.

2. Pour the lemon over the kale and gently massage until the kale begins to wilt. Set aside for an hour for the lemon to do its work.

3. Pour the olive oil over the kale and season with the salt. Add the grated carrots, golden beets, or sweet potato and gently toss.

4. Gently toss in the lemon zest, daikon radish, raisins, avocado, and sunflower seeds.

5. Taste and add drizzle of honey as needed.

Citrusy Root Vegetable Apple Slaw

How It Works—One of the best ways to utilize crucifers is in a slaw. Slaws are a great way to prepare crucifers, while incorporating many other Master Ingredients. They are also filling, quick to make, and—yes—nutritious. Traditionally (but not always), a slaw is thinly sliced cabbage mixed with another ingredient (such as grated carrot) and then tossed with a dressing. You can make a slaw from cabbage, but you can also make it from kohlrabi or turnips. You can use all three of these ingredients and even add in thinly sliced kale. Additional ingredients add color and additional nutrient potency. A raw slaw preserves the critical sulforaphane bioactive. Take it to the next level by finely slicing milder crucifers such as bok choy into salads and slaws.

(Personally, I make slaws with a variety of different ingredients such as raw beets and raw sweet potatoes. I use them as the base of the slaw and not cabbage. For that bioactive-gene influence and color, I will add in raw, thinly sliced kale or maybe bok choy.) There are a myriad of slaw recipes on the internet. Go ahead and search (or check our Genomic Kitchen Pinterest boards), and make a few slaws that you like. To change things up, I like the Root Vegetable Slaw below adapted from a *Moosewood Cookbook* recipe. The sweet-tart taste of this slaw makes the more robust root vegetables approachable. And instead of cabbage to kick in the sulforaphane, we use arugula.

Or you can throw in finely slivered kale or bok choy leaves right at the end. Easy!

PREP: 30 MINUTES (20 MINUTES IF YOU USE A FOOD PROCESSOR TO GRATE). SERVES 4.

GRAB THESE

2 cups peeled and grated tart apple (approx. 3 apples)

2½ cups grated root vegetable (example: celeriac, rutabaga, turnip, golden beet)

Juice and zest of 1 lemon

Juice and zest of ½ large orange

1 tablespoon olive oil

¼ teaspoon salt

Pinch cayenne

⅓ cup slivered fresh basil, or 2 tablespoons slivered mint

Generous handful of fresh arugula, or finely chopped kale, or leaves of bok choy

½ cup raisins or golden raisins

Drizzle of honey or maple syrup, optional

DO THIS

1. As soon as you have grated the apple and root vegetables, toss with the lemon zest and juice to prevent browning.

2. Stir in the orange juice and zest, olive oil, salt, and cayenne. Coat the ingredients well to distribute the flavors.

3. Just prior to serving, gently fold in the herbs, greens, and raisins.

4. Do a quick taste test. If too tart, add a drizzle of honey or maple syrup to punch in some sweetness.

Cooking Crucifers and Adding a Bioactive Punch Back In

Curried Lentils with Cauliflower and Greens

How It Works—You now know that applying heat to crucifers denatures the important myrosinase enzyme. Do that and you can't create the equally important sulforaphane bioactive. A little secret here: scientists have discovered that bacteria can actually produce sulforaphane in the gut. But so many of us have poorly functioning guts, so the likelihood of this happening is slim. Back to our recipe. A quick look at the ingredients and you see that we are cooking a vital crucifer, cauliflower. Yep, we are denaturing our enzyme. But not to worry, right at the end of the recipe we stir fresh greens back in. If you use spinach or chard, you don't get sulforaphane! But if you choose wisely and use a leafy crucifer, you do add the magic back in. This is a great recipe to use *fresh* (check the date!) pre-packaged baby kale if you don't have fresh whole leaves on hand.

PREP: 20 MINUTES. COOK: 40 MINUTES

SERVES 4 HEARTY PORTIONS.

GRAB THESE

1 cup green or brown lentils, rinsed and drained

2¾ cups water

2 teaspoons olive oil

1 onion, diced

1 clove garlic, minced

1 teaspoon sea salt, or to taste

1 teaspoon each coriander, ground cumin, and turmeric (or use a curry paste)

½ teaspoon ground cinnamon

2 cups cauliflower florets

1 cup tomato sauce

1 teaspoon ginger, dried or 2 teaspoons fresh grated

Pinch cayenne pepper flakes

2+ cups or 1 small bunch spinach, baby kale, or kale mix rinsed

Roasted cashews for garnish (optional)

DO THIS

1. Bring lentils and 2 cups water to boil. Lower heat, cover and simmer until tender and water absorbed, 25-30 minutes.

2. Meanwhile, in a separate pan, heat olive oil and sauté onion, garlic, and salt until onion is soft, not browned.

3. Add all remaining ingredients except the greens and nuts, plus ¾ cup water. Cover and simmer until cauliflower is *al dente*, +/-15 minutes. Add more water as needed to keep the sauce fluid, but not runny.

4. Stir in the greens. Cover pot and allow to wilt for 30 seconds to 1 minute.

5. Gently fold the cauliflower mixture into the lentils.

6. Garnish with cashews.

Chapter Seven: Influencer Ingredients: Nuts, Seeds, Animals, and the "Bs"

The Influencer Ingredients are, you guessed it, influential! When I first wrote this chapter, I attempted to stuff everything I could tell you into it. The chapter grew and grew. And grew and grew. Fifty pages later, before I'd even gotten into the kitchen, I finally realized that even the greatest of information processors among us would get lost. So, I am telling the story of our Influencers in three chapters. In this chapter we will focus on the role of a segment of our Influencers in important biochemical processes in the body. I call these the Nuts, Seeds, Animals, and the "Bs"—B vitamins. In the second of our Influencer chapters, I will address omega-3 fats and why they are themselves deemed "influential." In the third and final part, Influencers in the Kitchen, you can join me in cooking with them. Let's get started.

The Science Behind Influencer Ingredients

Your body is an advanced, complex biological machine, like a car that has many independent functioning parts that help it to run smoothly. You car's parts are organized into systems: the wires and cables and fuses of the electrical system make the lights shine and spark the cylinders; the power train's transmission and gears harness the firing of those cylinders to turn the wheels; the emissions system's exhaust pipes and catalytic converter collect and evacuate the engine's waste products. Similarly, your body has many processes and cycles that need to be properly maintained for us to live a healthy, active life.

What controls how those many parts interact and function in the car is its computer; in your body, each cell is a functioning manufacturing plant. Your genes are integral to the day-to-day operations of the biochemical complexes that make up your body. Let's take a closer look at some of these biochemical pathways and how they impact your health, so that you know why our Influencer Ingredients are indeed influential.

Recap on Genes

As you know, our body houses trillions of cells. Each individual cell has a nucleus that contains our genes which are made up of DNA, the genetic material we inherited from our parents. DNA contains the instructional material for making proteins which are the movers and shakers in our body. So the question is: how do we get genes to start and then stop making proteins? Let's review this process first so that we can understand why we need Influencer Ingredients.

We need to turn genes on so they can produce proteins necessary for our bodies' everyday functioning. But, like many things in life, it is a complicated process. Since this is not a book about molecular biology and advanced genetics, we'll keep this simple. The cells in your body continuously respond to, and react to, signals that turn genes on and off. The act of turning genes on or off is called *gene regulation*. Genes are turned on and off in the body's response to achieve balance, or homeostasis. Genes may be turned on in reaction to a need to grow more cells, for example growing a baby or responding to injury. They may also be turned on to respond to different foods that you are eating. Hormones, signals from the environment, and interactions with other cells are examples of how genes get switched on and off. Genes are therefore continuously activated or activated to manufacture more proteins to respond to a specific need the body has. When that need has been met, the manufacturing process must be curtailed.

One principal means by which genes are activated we learned about previously is acetylation and deacetylation. Another means is via a transcription factor, which you have already read about in previous chapters. Bioactives in our food are one way that we can nudge the transcription-starter motor into action, which in turn starts the DNA production wheel turning. Remember Nrf2? Nrf2 is a transcription factor that nudges specific genes to hand off their DNA instruction. We talked about how bioactives work with NrF2 in Chapter 5, Master Ingredients.

Turning Genes Off

Your body is in constant motion, striking a fine balance between protein production and halting the protein manufacturing line. In humans, there are many mechanisms that can influence your genes to stop production. In this book, we'll focus on a process called DNA methylation. It is perhaps the best studied mechanism and a little easier to understand than other processes such as histone modification and chromatin remodeling. And yes, we did talk about this in passing when we discussed SIRT-1, but here we'll slow things down a bit and give you a better peek inside the methylation process. This will help you appreciate the role of our Influencer Ingredients a little more.

Think of your genes like a book. When you open a book (or your tablet), you can read words which form a story. When you close the book, the story ends until the next time you open it. For genes to produce proteins, their DNA code has to be open or exposed so that it can be read, or transcribed. Once sufficient proteins have been produced, the protein manufacturing process must be halted. Methylation is *one* way that we can turn off genes. In the methylation process, methyl groups which comprise one hydrogen atom with three carbons fixed to it are attached to a specific area of your DNA. (For the curious among you, this equates to the C5 position of the cytosine nucleotide and the fixing of the methyl group that forms 5-methylcytosine.) Now remember the transcription factor? Transcription factors nudge DNA into action, advising the nucleotides to start handing out a protein recipe to RNA. When methyl groups attach to DNA, they prevent the transcription factor from engaging with DNA. In other words, DNA methylation shuts the cookbook so protein recipes can't be read or made. Methylation is one mechanism by which the cell can control gene expression, depending on the body's needs. Your body is in constant biochemical motion and it needs can change from one second to the next.

Okay, now that we understand one mechanism for turning genes off, the question we need to revisit is why and when do we turn genes off? You know already that genes are activated to replicate or generate cells to respond to periods of growth such as in the womb, during adolescence, or when recovering from trauma or surgery. The same process of growth applies insidiously to cancer where carcinogenic cells replicate out of control. In this scenario, we want to block cell growth or replication. Methylation is a process by which we can lock genes in the off position, thereby shutting down cell replication and growth activity. Unfortunately, mishaps in the methylation production process, which can be caused by gene SNPs, environmental exposures, or nutrient deficits, can influence whether genes are correctly shut down or not.

This leads us to the "methylation cycle." From its name, you can probably deduce that the methylation cycle generates the methyl groups that can silence, or deactivate genes. But methyl groups have another function, namely modifying or adapting compounds that are essential to so many biochemical processes in the body. Let's learn a little more about the methylation cycle and why it is instrumental to your health.

The Methylation Cycle: Controller of Controllers

While your genes are instrumental to so many different bodily systems, there is one system which is, you might say, the "controller of the controllers." (FYI, the full name of this is the transmethylation cycle, and we'll explore that just below.) The methylation cycle functions somewhat like traffic circles, which are designed to keep traffic moving smoothly. Instead of cars piling up as they sit at a stop light, the circle allows traffic from all directions to enter and exit the flow smoothly. No one has to stop, and with a bit of driver courtesy, traffic comes and goes without a backup.

The methylation cycle helps to move biochemical traffic from the center of the city to the 'biochemical suburbs' of your body. It does this by moving biochemical compounds in a cycle, modifying them, and then sending them into the right traffic lanes and adjacent cycles where they support important functions in your body.

Transmethylation: A Closer Look

Remember, the full name of the cycle we're examining is transmethylation. Let's break down the word. "Trans" means to move or transfer. Methylation refers to a "methyl group" or the biological unit of one carbon atom surrounded by three hydrogen atoms. The fundamental function of this cycle is to move or transfer these biological units between compounds in the cycle. Now the methylation cycle does not operate in isolation. Like old-time clocks which are comprised of a series of wheels whose cogs (teeth) interlink, one turning the other, the methylation cycle interacts with other cycles. Compounds produced by the methylation cycle are handed off to other biochemical cycles which together form your physiological operating system. I call this interlinking of biochemical cycles our *inner biological circuitry*. Methylation is at the heart of how the cycles turn, interact with each other, and work together to support our health. If the methylation cycle functions efficiently, we hum along in full health. If it does not work efficiently, you can imagine a "knock on" effect similar to a line of falling dominos or a poorly-tuned engine or wheels whose cogs fail to mesh. When one part of the cycle gets out of round, all associated cycles are affected. That being the case, an efficient cycle depends on two things:

- That we fuel the cycle with the nutrients it needs to operate effectively; and

- That there are no SNPs or gene variants in the cycle that prevent it from operating effectively.

Know Thy Neighbor

To understand the importance of the methylation cycle, let's use the example of a neighbor, the trans-sulfuration cycle. Like the methylation cycle, the trans-sulfuration cycle produces many compounds that are essential to your health. A notable one is glutathione, instrumental in detoxification, the process of removing toxins from the body. The trans-sulfuration pathway also produces the amino acids cysteine and taurine, both with specific operating functions in the body. The methylation cycle must work efficiently to provide the "starter" biochemical compound for the trans-sulfuration cycle. Both cycles work together like cogs in a wheel. Thus, for both cycles to operate effectively, we need to ensure that the core nutrients are in place. I've listed these nutrients below. And don't worry, I'll show you the *foods* you need to include in your pantry to support these cycles. No need to memorize nutrients!

Influencer Ingredients contain the nutrients to support the methylation cycle itself, and the cycles it directly links with. Take a quick look at the list below, and you'll see that the B vitamin family is well represented, as so many of the Influencer Ingredients are rich in different B vitamins.

Essential Nutrients:

Vitamins: B2, B3, B6, B12, Folate

Minerals: magnesium, selenium, zinc

Amino Acids (protein building blocks): cysteine*, glutamate*, glycine*, methionine

Other: choline, betaine

* Note: cysteine, glutamate, and glycine can all be made in the body, providing all the essential nutrients are in place to enable their production. For example, cysteine is produced in the trans-sulfuration cycle.

Now that you understand that the methylation cycle is integral to how biochemical traffic moves around the body, let's look at just a sampling of the processes that methylation and the methylation cycle influence; you'll quickly see why it is so important.

Brain Health

The BH4 (Tetrahydrobiopterin) cycle forms important neurotrans-mitters or "neurotalkers" that influence brain chemistry and your mood. The methylation cycle interacts with the folate metabolism cycle that in turn, interacts with the BH4 cycle. Each interaction produces compounds. In the case of the BH4 cycle, we're talking about neurotransmitters, or hormones like epinephrine, norepinephrine, dopamine, serotonin, and melatonin.

Epinephrine and norepinephrine: These are our "fight or flight" hormones which are released by the body under stress. Dopamine, serotonin, and melatonin are our feel-good hormones. All are produced by the interaction of the BH4 and folate metabolism cycles. You can see that if one cycle fails to communicate effectively with another, some of these important compounds may not be produced, and you may feel stressed, agitated, moody, or perhaps euphoric! Now back to fight or flight for a moment. After the adrenaline hits, we need to get rid of the excess chemicals (epinephrine and norepinephrine), or we'll feel permanently on edge, similar to when you have ingested too much caffeine. This is where the methylation cycle comes back in. To properly dispose of epinephrine and norepinephrine, the brain requires a specific enzyme called C-O-methyltransferase or COMT. COMT grabs a *methyl group* and attaches it to the neurotransmitter, thereby rendering it less toxic and more user-friendly for the body to handle. The methyl group comes from guess where? The methylation cycle. Without that methyl group, COMT cannot do its job, and you might find yourself walking around in the jittery "on" position. You see, *everything* is interlinked! Are you starting to see just how important that methylation cycle is and why we ensure you have the nutrients you need in our GK Ingredient Toolbox to provide the support it needs?

Detoxification: This complex process removes toxic chemicals, heavy metals, and by-products of hormones from the body. Glutathione is manufactured in the trans-sulfuration pathway, fed directly from the trans-methylation pathway. Glutathione is a master ringleader in your detoxification system and requires glutathione peroxidase (encoded by Nrf2) and selenium (we put ingredients sources in your toolbox), to operate.

Energy: We're all interested in energy and making sure we have plenty of it! Energy fuels how our body works in the form of ATP or Adenosine Triphosphate. Making ATP is a complex process, but suffice to say that a single compound found in every cell of your body, CoQ10, is instrumental in the final stages of how this high energy molecule is produced. Just as the enzyme COMT needs a methyl group to do its work, so a methyl group is required to manufacture CoQ10. Talking about making energy, carnitine is responsible for transporting fatty acids across the mitochondrial membrane (the mitochondria is where ATP is made), so that they can be processed into energy. Building block proteins for carnitine require methylation in order to produce the carnitine molecule. Are you starting to see just how important the methylation cycle is?

Creating New DNA: Purines and pyrimidines are the compounds from which your DNA derives the base nucleotides, or basic units of which it is built: adenine, guanine, thymine, cytosine. Purines and pyrimidines are downstream products from the methylation cycle. If you can't generate new DNA, you can't create new cells to replace old ones. This is particularly important for your immune system. A slowdown in DNA synthesis might impact how fast your body can build up the cellular defenses of your immune system.

Your Genes and Cancer

The trans-methylation cycle and its neighbors like the trans-sulfuration pathway are pretty important to how you function. You can see from our list above that they impact many processes, including DNA production itself. Spelling errors or SNPs in genes producing the enzymes that help the methylation cycle function may have very detrimental impacts on your health, depending on which ones they are and how many you have in the cycle. If the cycle is not working effectively, we are not producing enough methyl groups. Or in some cases, we can produce too many. Let's see what the impact of this might be as it relates to cancer.

In the ideal scenario, we want to turn *off* cancer-causing genes, and we want to turn *on* tumor-suppressing genes. We can do this by producing methyl groups in the right quantity. Unfortunately, when the trans-methylation cycle is not working efficiently, we may not produce sufficient methyl groups to *turn off* the cancer genes, a situation called global hypo-methylation. Conversely if the cycle is running amok and we produce too many methyl groups, we overwhelm the tumor suppressing genes by *hyper*-methylating them, which renders them useless. If that happens, they can't suppress tumors, and the cancer progresses.

How do we prevent this situation? We make sure that we have all the core nutrients in place to support the efficiency of the cycle. In cases where we have SNPs or variants in the genes that support the cycle, we can measure levels of nutrients needed for the cycle, and also compounds (we call them metabolic intermediaries in the clinical business) produced in the cycle. If we find sub-optimal levels for either, we are able to intervene and restore them to normal levels using food and/or supplementation.

Let me give you an example. Folate (vitamin B9) and vitamin B12 are what I call gateway vitamins for the methylation cycle. These two vitamins work together at the entry point into the cycle. If either of these vitamins is deficient, or even scarce in your diet, the methylation cycle will struggle along because the starter motor for the cycle provided by folate and vitamin B12 is not functioning well. My advice to you is to seek the help of a doctor or nutrition expert to evaluate how well your methylation cycle is working, and please don't tinker on your own!

OK, OK, Enough Science. Can We Get to the Food, Please!

You have had a huge dose of science I know, but as I always say, nutrition advice without culinary translation is just a bunch of advice! If we don't translate science to the plate, we have an empty plate. Collectively, Influencer Ingredients contain all the nutrients you need to support the important methylation cycle and its immediate neighbors like the trans-sulfuration cycle and the folate metabolism cycle. In fact, if you combine them with your Master Ingredients, you have most (but not all) of what you need to help your body beneficially navigate its biochemical highways. The graphic on page 130 shows you what the Influencer Ingredients are.

For this Influencer Ingredients chapter, let's focus on all the Influencer food groups except for *Seafood*, which we'll address in next chapter, Influencers Part 2. If you compare Influencer Ingredients to Master Ingredients from Chapter Five and Six, you'll see I have made a few additions to the fruit and vegetable sections, but importantly I have added an *Animal Protein* section and an equally important *Nuts and Seeds* section. In the *Other* section, you'll find yeast and two grains. The added ingredients are rich sources of B vitamins, zinc, magnesium, choline, betaine and the amino acid, methionine. All of these nutrients are essential to smooth operation of the methylation cycle, which is why we've packed them into your GK Ingredient Toolbox.

Going a bit deeper into why I have included specific ingredients, the nuts, seeds, and listed grains were chosen because all are excellent sources of essential minerals as well as the vital B vitamins we've been discussing. Animal proteins are by far the most robust source of methionine which is why they are added to the GK Ingredient Toolbox. I cannot emphasize just how important methionine is to the methylation cycle. If you do not eat animal protein, I ask you to be sure you include Brazil nuts and spirulina (blue-green algae available as a dietary supplement) on your daily plate as they supply methionine, though in minimal quantities compared to animal proteins.

Eggs and beets are added for a very specific reason. They are both a good source of a compound called trimethylglycine (TMG). Some of us have problems in the methylation cycle converting one of its required intermediate substances, homocysteine, to methionine. This conversion requires folate and B12, and can be derailed if either vitamin is unavailable in sufficient quantities, or not in the right form to perform the conversion process. SNPs, by the way, can cause this aberration. But the body is clever! In biochemistry, there is a side door through which homocysteine can be converted to methionine, subsequently circumventing the B12 or folate problem. Trimethylgycine, known as betaine in beets and spinach, and produced from choline in egg yolks, is our ace in the hole, providing us with an alternative work-around to keep the methylation cycle moving smoothly. Time to get busy with beets,

INFLUENCER INGREDIENTS

Fruit	Seafood	Animal Protein	Legumes
Avocados	Albacore Tuna	Cheese (Aged	Black-Eyed Peas
Mango	Anchovy	Preferential)	Soybeans*
Oranges	Herring	Meat	
Sun Dried Tomatoes	Mackerel	Poultry	**Other**
	Mussels	Whole Eggs	Brewer's Yeast
Vegetables	Oysters		Mushrooms (Shitake)
Beets	Pacific Halibut	**Nuts & Seeds**	Nutritional Yeast
Corn	Salmon	Chia	Quinoa
Peas	Sardines	Brazil Nuts	Rye
Peppers		Flaxseed*	Sea Vegetables*
Spinach		Hemp Seed	Wheatgerm
		Sesame Seed*	
		Sunflower Seed*	
		Walnuts	

Asterisks indicate ingredients that appear in other MISE categories.

which we'll do in Chapter Nine, Influencers in the Kitchen, when you get there!

OK, now that we have reviewed these Influencer Ingredients, let's head over to the omega-3s and dig into why I have included them in our GK Ingredient Toolbox. See you in the next chapter.

Chapter Eight: Influencer Ingredients: Omegas and Other Fats

Now that we are through our nuts, seeds, animals, and "B's," let me turn your attention to fat. Yes, indeed, that much aligned nutrient that has been hidden, avoided, denigrated, and banished to the lowest rung of the nutrient ladder for its "destructive" nature. At least that was the pattern of thought for a couple of decades. In this chapter, I would like to reshape your thinking a little bit with a focus on why fat is important, and where—particularly—we need to (re)focus your attention. Let's get started.

From Facts to Fats: Better Biochemistry through the Right Fuel

My foray into food began as a kid and deeply influenced my palate. Veggies from the garden. Full fat milk on the doorstep (I grew up in the UK when milk was delivered in the morning with a fat creamy top.) Yellow butter. Foraged fruit from the hedgerows. Memories aside, my training in nutrition took place in another, later era when the thinking proved to be the antithesis of my full-fat childhood. Enter the era of low-to-no fat with added sugar substitutions. Yep, I myself probably taught people how to dump the fat, increase the (whole) carbs, and, unknowingly, played a part in spurring the evolution of obesity. Dumping the fat and increasing the carbs (albeit not sugar), was what the science said, after all. Yet the everyday common sense of my European roots reminded me that synthetic food minus the fat was not in keeping with nature's way and surely did not belong on the plate. I absolutely knew that way before we understood the benefits of the much-touted Mediterranean Diet. The long-lived folks from those countries surely were not replacing local greens, full-fat cheeses, and lashings of olive oil with low-fat alternatives. In fact, they couldn't be bothered because low fat was an American phenomenon and was found in supermarkets and not at their local markets, in their gardens, or at the local *taverna* or *trattoria*.

A few words about fat. You need fat in your diet. Period. Saturated and unsaturated fats are integral to the structure of your cell membranes, to hormones, neurotransmitters (those brain talking molecules). Fat is a leading communication tool.

In the low-fat era, we became obsessed with total fat and its sidekick, saturated fat. In fact we were so focused on getting anything fat-related under some magical percentage of total calories or whatever that the bigger picture of type and quality of fat was lost. We neatly organized fat into Total, Saturated, Monounsaturated (much heralded), and Polyunsaturated, and then got obsessed with counting calories to be sure that saturated fat never exceeded ten percent of total calories consumed. And as the saturated fats sped downwards in our diet, the presence of polyunsaturated fats blossomed as more and more products labeled "lower in saturated fat" appeared on our plates. Given that our body cannot make the omega-3 and -6 versions of polyunsaturated fatty acids, and that saturated fats were deemed heinous, the low-fat era allowed us to get a little more acquainted with polyunsaturated fats, in addition to our full embrace of copious carbs. It is certainly a good thing to branch out by varying the *kinds* of fat you consume, replacing saturated fat with more polys, for example, and our thinking around fats has evolved a great deal since those days. Even so, there's a catch that can come back to bite us if we aren't careful. You see, the devil is always in the details.

The Devil is in the Details

Because the body is not capable of making omega-3 and -6 versions of polyunsaturated fats, you know that we have to source them from our diet. The omega-3 and -6 versions come with a paradox, or maybe even "baggage." They both need to be present for best use in the body—but in the right ratio. That ratio counts for everything, as you're about to find out.

Our two lead omegas work synergistically with each other when they are eaten in the correct proportions. The right ratio is 4:1, or four units of omega-6s to every one unit of omega-3. Things go haywire when you have too many food sources containing omega-6 and not enough omega-3. And in the age of convenience foods and eating out, it is very easy to fuel this dysfunction since there are so many more sources of omega-6's out there. If you have any packaged foods at home—snacks, crackers, frozen pizza, or desserts—take a look at the ingredient label. Sift through the ingredients until you get to an oil—generally soybean, safflower, sunflower, corn, cottonseed, peanut, though these days maybe you'll find canola, palm, or coconut, too. Now if you take a drive across America's heartland, which I did in the fall of 2017, you actually see many of the plant sources of these fats growing outside your window as you pass by. Thousands and thousands of acres of sunflowers stand with their alert heads tracking the sun. In another direction, field after field of delicate canola flowers wave in the wind. Elsewhere, the green of soybeans and arching stalks of corn. Everywhere you look, fields of green and yellow growing oil for your (microwave) dinner! The reason I mention this is because while these plant sources of oil are a prevalent source of essential omega-6 fatty acids, they contain nary a molecule of necessary omega-3s.

As you read those labels, you might also see the acronyms "EPA" and "DHA" which often come up when we talk about the omegas. These acronyms stand for eicosapentaenoic acid and docosahexaenoic acid, two important essential omega-3 fatty acids. Even though you'll learn about them both in detail later on, suffice to say for the moment that, while omega-6s and -3s are widely available in various foods we consume, they are not necessarily available in the correct chemical *form* the body can use, which is why DHA and EPA are so important. They are available as an immediately useful form and have the greatest health benefits.

Your body has to work harder for your omega-3 fats because they are tricky little devils. They are found most robustly—in most user-friendly form to the body—in specific seafoods like wild salmon, halibut, mackerel, sardines, and scallops, to name a few. They are also in animal products—meat, dairy, eggs—which are free-range and/or grass-fed. You've probably read that some nuts, seeds, and some vegetables are sources of omega-3s as well. This is true, but again, there is a twist here. Before we uncover the backstory of plant sources of omega-3s, let's talk more about the benefits of these fatty acids and why we need to eat them in the right proportions.

Ratios and Benefits without the Math

Our ancestors obtained important fats through their diets, mostly from the animal protein they hunted and some from the plants they foraged. They did not belly up to a bowl of (extruded) cereal in the morning, or pop cinnamon rolls or a pizza in the oven *ever*. Now, we like to think they had an idyllic process-free food life, but frankly, many of them were probably lucky to get a regular meal. They lurched from feast to famine. The season (and climate) dictated what was available, and we can be pretty sure that they had no interest in getting lots of colors on their plate, counting calories, or wondering whether they had reached their recommended daily intake of any given vitamin or mineral. They were simply surviving, with the occasional lucky family possibly thriving. But, a common thread through all of this is that the ratio of omega-6 polyunsaturated fats to omega-3s was far more in balance than today. Researchers believe that our ancestors (and maybe even your grandparents, depending on how old you are), had a ratio of omega-6s to omega-3s in their diet of approximately 4:1. Why? Because omega-6's and omega-3's found in natural—unprocessed—foods (animal proteins, seafoods, some nuts, seeds, plants) occur in the 4:1 ratio conducive to our health.

Today, as a general rule of thumb, if your diet includes a lot of packaged foods, chances are you are eating a lot of omega-6s in the form of oils extracted from soybeans or seeds, such as safflower and sunflower and relatively fewer omega-3s. So, your ratio is off. And I am not just talking about snack foods or frozen foods, either. If you are eating out, many of the foods served in restaurants are not made from scratch, but rather are partially processed and combined or reheated before serving. You will never know this because your dinner plate at the diner does not always come with a full ingredient disclosure label.

The abundance of omega-6 tilts the balance from our ancestor's healthy ratio of 4:1 to an estimated 22:1, i.e. 22 omega-6s to 1 omega-3. Imbalance like that is never without consequence. Let's be very clear however that we *need* both omega-6 and omega-3 for optimal health. Whether you are eating these nutrients from plants, oils, nuts, seeds, seafood, beef, chicken, eggs, or spirulina, you need both fat sources. Sunflower oil is not the devil. Wild-caught salmon is not the bee's knees. Canola and olive oil contain both 6s and 3s. Nature isn't stupid! The key is that we need both sources, but in balance. Let's see what happens when our omega ratio gets out of balance.

Bring on the Pain Pathway

If any of you have ever studied the biochemistry of fat metabolism, you know that it is as fascinating as it is complex. And perhaps most fascinating of all is how the body handles its omegas. In nutrition, we are largely taking complex molecules and reducing them into simpler parts so that the body can assimilate them. The omegas work a bit differently and, depending on their source, actually have to become a little bit more complex before they can be user friendly to the body.

Metabolizing the omegas is like some kinds of team competition. The more folks you have on the field, the more likely you are to beat your opponent because you simply have more bodies to act. The same applies to our omegas! With one major exception, the omegas are in a competition with each other, and omega-6 will usually win.

As we learned earlier, in order for the body to utilize omega fats, they have to be in forms the body can use. In the case of omega-3 fats, those most user-friendly forms are the long-chain fatty acids DHA, DPA (docosapentaenoic acid), and EPA. In the case of omega-6 fats, those forms are the shorter chain fatty acids

CHAPTER EIGHT: INFLUENCER INGREDIENTS: OMEGAS AND OTHER FATS

GLA, DGLA, and AA (gamma linolenic acid, dihomo-γ-linolenic acid, and arachadonic acid, respectively). To get the most out of omega-6 forms, the body must convert, or elongate, the original source of each fat to its more complex (long chain) user-friendly form! Alpha-linoleic acid is the starting compound associated with omega-6 fats, and alpha-linolenic acid is the starting compound associated with omega-3 fats, with a major exception. Omega-3 fats derived from certain seafood, grass-fed animal products, and spirulina (DHA form) are already in their user-friendly EPA-DHA form and don't need to be converted by the body. All other foods with "trace amounts" of omega-3 need to be converted because the fatty acids are in those ALA "starter" forms and not in their user-friendly EPA-DHA forms. Ah ha! More on this later. (Omega geeks: click here to see a graphic tracing the conversion steps for Omega-6 and -3 fats.)

Going back to our conversion process, it gets tricky. Think of both omegas like two ice hockey teams. Equal players on each team with equal opportunities to shoot at the goal. Now a penalty is called on one of the teams, and they are down a player. There is an imbalance on the ice and a competitive advantage to the team with all players. Imagine a second penalty on the same team, and they lose another player. Now there is a real competitive advantage, particularly the longer the situation endures.

In our omega scenario, right off the bat, in the first step of their conversion, there is a competition for the same nutrients and the same enzyme that converts both omega-6 and simple omega-3s to their more complex user-friendly forms. Which side do you think will win? Who's got more players in the game? Like our ice hockey team, the side with the most players on the field wins. In our time and modern diet, we already know that omega-6 is ubiquitous in packaged and processed foods, so it will be more readily consumed, and therefore have more players on the ice. It will win the conversion game. That's all well and good until you realize that winning the game can head the body down "the Pain Pathway" as we'll see below.

The Winner Does Not Always Take All

The omega-6 and -3 pathways are two biochemical pathways operating in balance with each other. One pathway allows us to bleed. The other to clot. One side stimulates a rapid immune response to wounds and trauma. The other helps smooth and soothe the rapid immune response, moving the body back to a calmer, less inflamed state. The omegas are like seesaws, where one balances the other. Just like on a seesaw, when three people climb on one end, one person dangles in the air on the other. Eat too many foods containing omega-6s—the natural tendency if we eat lots of packaged and processed foods—and we might fast track an inflammatory response, which is one mechanism associated with the omega-6 pathway. Under normal conditions, the body unleashes inflammatory responses so that you quickly remove your hand from a hot plate to prevent a burn. Or you cut your finger and experience bleeding, pain, and swelling. These are all automatic inflammatory responses induced by metabolites produced from the omega-6 pathway. Inflammation usually equates with pain, and pain is a signal from your body that something is wrong. Thus, I call the omega-6 pathway the Pain Pathway because it is the same pathway common pain medications like Tylenol, Aleve, and Ibuprofen work to block, specifically by stopping the signaling from the inflammatory response.

By consuming omega-6 fatty acids in more abundance than omega-3 fatty acids, you are, in effect, inadvertently engaging inflammatory responses that manifest in pain. What kind of pain? It could be as simple as a few aches or soreness. Or you might suffer headaches and irritability, or graduate to more complicated auto-immune disorders as the body grows weary of constantly fighting with an immune system set in the "on" position by the imbalance. As I tell people, one of the ways to tune down inflammation is to get more omega-3 rich foods on your plate. But remember, you *need* both of the omegas in a balanced ratio. No need to rush off and dump the omega-6s. They are important for your health, too.

The Importance of Omega-3 Fats: Tinkering Under the Hood Again

Over the years, you've probably read that omega-3 fats are nothing short of a miracle. They appear to be good for all of your body parts and systems. This broad-brush approach to nutrition has driven me more than a little nuts over the years. Take a look at how differently we talk about food in the science community versus in the chef community. In science, we'll grab beets and tell you what nutrient they are replete with and what body part that nutrient maps to—heart, kidney, bone, brain. Who cares? Nutrients are information for the body, and the body needs a variety of them.

Chefs, on the other hand, will grab a beet, tell you what it tastes like, what it pairs with it, and how to cook it to accentuate its flavor. In other words, they make a beet relevant and accessible to you and your palate. All this to say that I would like to remove nutrition mapped to body parts and systems and, once and for all, explain how omega-3s are powerful in your body in a more holistic sense. Here goes:

- Inflammation: Fats are an important part of the structure of all cell membranes.

- Cell membranes are an essential communication board for each cell, as well as a source of nutrients. Changes in the polyunsaturated fat composition of cell membranes influence the body's response to inflammation, skewing it towards a pro-inflammatory response if the cell membrane structure is richer in omega-6 fatty acids than omega-3. One of the ways omega-3s can mediate or block an inflammatory response is by interacting with master gene Nf-kB. By removing Nf-kB from the metabolic show, we slow down the inflammatory cascade that might otherwise occur.

- Heart Disease: what's the connection? First, the truly curious should seek out the brilliant work of Dr. Mark Houston, an integrative cardiologist. Here's the gist of what he is saying: When the structure of your vasculature (your veins and arteries) is undermined, it is called endothelial dysfunction. Dysfunctional or flabby blood vessels lead to a dysfunctional heart. Remove the insults that cause the injury, and you'll treat heart disease more effectively.

- Many things lead to dysfunction, inflammation being one of them. And one of the ways we trigger inflammatory responses is by toggling a series of receptors on cell membranes called Pattern Recognition Receptors (PRRs). Engaging these receptors sends signals to the cell, resulting in an inflammatory response. Activating the caveolae family (little flask-shaped structures in the cell also known as lipid-raft microdomains) has a similar impact. If turning these receptors on ignites a complex inflammatory response, then blocking them is surely kinder to our blood vessels and our hearts. Omega-3 fats are one nutrient we can use to block the receptors and downstream signaling injurious to our health.

- One of the things that most of us know about heart disease is that poorly controlled blood pressure often precedes it or accompanies it. Changes in blood pressure are often accompanied by changes in nitric oxide (NO). Nitric oxide helps to dilate blood vessels, ensuring the smooth flow of blood and nutrients. Block it, and the reverse happens! It turns out that when the caveolae receptors, mentioned above, are activated, nitric oxide production is diminished, causing blood vessels to constrict, and higher blood pressure to ensue. But our handy-dandy omega-3 is one means by which we can actually block receptor activation, thereby ensuring we have a ready supply of nitric oxide available so our blood vessels work smoothly. Now this is a *real* connection between omega-3 and heart health.

Smooth as Silk Blood Sugars

· I am always careful to advise people that there is no one miracle-worker nutrient! Sometimes it seems at any given moment that certain nutrients achieve Hollywood status while all others fade from the spotlight. As I write about the virtues of omega-3 here (and there are many virtues), remember that no nutrient works in isolation. All nutrient players are part of a team, even if it appears that one gets more attention than another. Diabetes and weight management are the forever "hot" topics, and we're all interested in them whether we are dealing with them or not. Which brings me to the PPAR family of genes we met in Chapter 5. This family acts as a master regulator of our metabolism. Omega-3 stimulates the PPAR family member which steers your cells to tap fats and burn them for energy. Good deal! Our handy omega-3 also gets the hormone adiponectin up and moving as well. Low levels of this hormone are associated with weight gain, obesity, and poorly regulated blood sugars. Why? Because when adiponectin is activated, it turns on the fat burning mode in your body and turns off the fat storage mode. When it is sluggish, you ramp up your insulin and start storing fat. This should get you thinking about the balance of your omegas if nothing else does.

The FADS-1 and FADS-2 Genes: Where SNP Siblings Can Lead to Trouble

So you now know that the user-friendly form of omega-3 fats is in the EPA-DHA form. You also heard that the body is able to convert a precursor form of omega-3 (alpha-linolenic acid) to EPA and DHA provided the right conditions are present. So it is possible to get our EPA and DHA from a variety of plant foods, which is where we find alpha-linolenic acid. Or can is it? This is where your genes come into play.

Converting alpha-linolenic acid into EPA and DHA takes a number of steps. Two of those steps involves enzymes called Fatty Acid Desaturases. If you have a SNP or spelling error on either the FADS1 or the FADS2 gene that encodes specific Fatty Acid Desaturase enzymes (an enzyme is a protein), then you may not be able to effectively convert your plant sources of alpha-linolenic acid into the EPA and DHA your body needs. FADS1 and FADS2 are instrumental to the individual steps in this conversion.

Essentially, if one of the conversion steps is faulty, it's akin to your body climbing a ladder only to find a rung is missing. In some cases, some of us can just take a giant step over the missing rung and keep climbing the ladder. For others, we just can't cross that distance.

Biochemically, there are consequences when your body can't climb that metaphorical ladder. If you are one of those people with either the FADS1, FADS2, or both SNPs, you *may* need some form of omega-3 supplementation (fish oil, for example) to bridge the gap, or you might need to consume it at the source, in the form of seafood and sea vegetables, like spirulina for example. How do you know if you have the SNP? Genomic testing will immediately tell you the answer. However, if that is not the route you want to take, ask your doctor to order an Essential Fatty Acids test. If your EPA or DHA levels are low, you know you need to bring on the omega-3 foods or consider supplementation. But whatever you do, be smart. Find out what your essential fatty acid levels are first, then add in food or supplementation and test your levels again.

Getting the Right Fats and Omegas on the Plate

Remember, you need fats in your diet because they are integral to the structure of your cell membranes, to hormones and neurotransmitters. Where confusion lies is in how much of each kind you need and what to take in or dump from your diet.

Here's what I know and what genomic research shows. One: read the first sentence of that last paragraph again. Two: differences in our genes can influence how our bodies package and utilize fat. This in turn may determine how much fat is necessary for you to eat and the proportion of saturated and polyunsaturated fats you should be consuming. Most of us do not have access to our personal genomic information currently, so we don't know how our genes are behaving. But we do have one clue that most of us get tested for at our yearly physical: triglycerides or cholesterol. If you have elevated levels of triglycerides or lower levels of HDL cholesterol, in spite of dietary prudence and exercise, I suggest you continue to work closely with your healthcare provider and consider some form of further, fairly common testing to look at how your body is handling fat. Such advanced tests, which may include lipid fractionation or specific genomic tests, will give you better clues about how you might want to tailor fat and nutrient recommendations to your situation. Regardless of tests, it's still a good idea to know how to optimize fat choices so we may ingest a cross-section from all fat categories. The graphic on the opposite page organizes and distributes optimal foods across both saturated and unsaturated fat categories. You'll notice that I include a mixture of animal and plant sources of fat. Fat, saturated or not, is widely distributed in nature across all kinds of food.

Quality Fat Chart

HERE ARE MY RECOMMENDATIONS WHEN IT COMES TO FATS:

- Make your fat choices as diverse as your produce choices by including a wide variety of foods on your weekly plate.

- You don't have to physically melt fat, pour fat, or spread fat on your food to get the benefits! Nuts and seeds are an easy go-to for some fat variety.

- Get serious about omega-3. You know the science. You know what they do. Make a concerted effort to include foods rich in the user-friendly, body-ready sources of omega-3 (EPA-DHA). These include specific seafood and products from animals that are grass-fed or grass-finished, the latter defined as animals never having consumed grains but eaten a foraged diet. If choosing animal products, I suggest buying these ingredients (when you can) from producers you know or can talk to.

- If you are not able to bring a grass-raised/finished product to your plate, or you are vegan, supplementation might be an important consideration. I strongly suggest you visit with your healthcare provider and request an Essential Fatty Acid profile to get a baseline measurement of your omega-3 and -6 levels. If both levels are optimal and in balance, it suggests that you are getting your omegas and converting the plant sources, too! If you have depressed levels of omega-3, first evaluate your diet to see if you are optimizing your omega-3 food choices. If you eat animal protein or seafood, you can

Making Quality Fat Choices from Natural Foods

BENEFICIAL SATURATED FATS	FOODS RICH IN OMEGA-6 POLYUNSATURATED FATS	MONOUNSATURATED FATS	FOODS RICH IN OMEGA-3 FATS POLYUNSATURATED
PLANT SOURCES	Animal	Most Nuts	Eggs, Meats
Plant Based Oils AVOCADO OIL, COCONUT OIL, MACADAMIA OIL, OLIVE OIL	BEST CHOICE: PASTURE-RAISED/FINISHED OR WILD GAME. PRODUCTS DERIVED FROM ANIMALS FED OR RAISED THIS WAY	Avocados, Olives Oils AVOCADO OIL, OLIVE OIL, PEANUT OIL, HIGH OLEIC SAFFLOWER AND SUNFLOWER OILS	Seed Oils LIMIT AND MIX WITH OILS FROM OTHER FAT SOURCES
Fruit ALMONDS, COCONUT MEAT, CASHEWS, HAZELNUTS, PECANS, SESAME SEEDS, SUNFLOWER SEEDS, WALNUTS	Seafood ANCHOVIES, HERRING, MACKEREL,OYSTERS, SARDINES, SALMON (WILD), TUNA		Nuts/Seeds ALMONDS, BRAZIL, MACADAMIA, WALNUTS, HEMP, PUMPKIN, SUNFLOWER
	Vegetables ARUGULA, CAULIFLOWER, DARK LEAFY GREENS, PURSLANE,		Legumes PEANUTS, SOY
PRODUCTS FROM GRASS-FED ANIMALS	RADISHES, SEAWEED, SPINACH, SUMMER SQUASH, WINTER		Fruit AVOCADO, OLIVE
Meat & Poultry	SQUASH		
Dairy BUTTER, GHEE, MILK-BASED PRODUCTS (CHOOSE CULTURED)	Fruit AVOCADO, OLIVE		Supplements EVENING PRIMROSE, COD LIVER OIL, FISH OIL, BLACKCURRANT OIL
	Herbs/Spices BASIL, CLOVES, MARJORAM, OREGANO		
Eggs			
	Legumes KIDNEY, NAVY BEANS		
	Nuts & Seeds BRAZIL, CHIA, FLAX, MACADAMIA, PINE, PUMPKIN, SUNFLOWER, WALNUTS		

zero in on the best protein sources available. If you are vegan, you need to hone in on foods that are the best source of alpha-linolenic acid and focus your attention there for a few weeks. Then repeat your lab test. If your levels increase, you know you can achieve your fatty-acid balance through diet. If your levels do not respond, you do need to consider supplementation.

Where to Find Your Omegas and Other Fats You Need

Fat is all about balance and quality. By balance I mean mixing up your saturated fats with their mono- and polyunsaturated siblings. Not too much of one or the other, but a veritable compendium is what we are after. Use the Quality Fat Chart above to help you navigate a variety of different fats from various sources onto your plate. Be sure to choose fat from each category and not just one. Remember, you *need* fat from a variety of sources because different fats play different roles. Not too much. Not too little. Think variety. Think quality. It's not about mathematics. It's about variety.

Oil is Like a Precious Metal—Except That It Can Lose Its Luster With Age

Let me leave this chapter with a further, final note for you, and a deeper thought about fat and quality. In the fall of 2017, I was hosting a program in Florence, Italy. One of the planned excursions was to an olive oil producer—after all what's a trip to Tuscany without tasting olive oil? One of the beautiful things about this producer was his insistence on date stamping his product. Some of his product is so young and so pure that only the locals could really get it because its virgin structure meant it could not survive the rigors of packaging and extended transportation. No problem there because, as good Italians, the locals tend to cart around their olive oil in 3-liter jugs or tin containers rather than the pretty bottles we see here at the fancy food stores. I vividly remember a conversation between the producer and a chef who was accompanying our tour. They were laughing about the sheer quantity of olive oil they consumed monthly. I think I overheard "at least three liters."

Today, this particular producer exports his own product, untainted by blending with oil from other producers, to the USA. Everything is date stamped. He knows his product, and he knows the detriment of heat and light to the quality of his oil. Heat and light denature the product, causing off-flavors as the fat slowly oxidizes. The product one year later is not the same as the day it was bottled. He told us emphatically that he would rather pull all the product off the shelves than sell it one year after its production. Wow, I thought a product with a promise and probably a money-back guarantee! You see olive oil expires and loses its aromatic flavor and nutritional structure. As one California olive oil producer notes, "unlike wine, olive oil does not improve with age!" All this to say that olive oil is the nutritional equivalent of a precious metal. Treat it like a diamond and don't let its value deteriorate. The same applies to all oils, including flax oil. Look for a best-by date, but better still a production date. Then you actually know how old the product is. Buy it in a dark bottle and keep it in a dark and cool place. Use it frequently. Get rid of the old stuff, meaning if it is older than 2 years, or preferably older than 18 months, dump it. Listen, you should be *using* oil, not admiring it. And last but by-no-means least, if you open your bottle of oil and it smells like paint thinner, dispose of it like a paint thinner. After all, you don't drink paint thinner, do you?

And on that note, let's wrap up with another look at the complete set of Influencer Ingredients we're going to be working with. Now let's head over to the kitchen and get them onto your plate. See you there.

INFLUENCER INGREDIENTS

Fruit	Seafood	Animal Protein	Legumes
Avocados	Albacore Tuna	Cheese (Aged	Black-Eyed Peas
Mango	Anchovy	Preferential)	Soybeans*
Oranges	Herring	Meat	
Sun Dried Tomatoes	Mackerel	Poultry	**Other**
	Mussels	Whole Eggs	Brewer's Yeast
Vegetables	Oysters		Mushrooms (Shitake)
Beets	Pacific Halibut	**Nuts & Seeds**	Nutritional Yeast
Corn	Salmon	Chia	Quinoa
Peas	Sardines	Brazil Nuts	Rye
Peppers		Flaxseed*	Sea Vegetables*
Spinach		Hemp Seed	Wheatgerm
		Sesame Seed*	
		Sunflower Seed*	
		Walnuts	

Asterisks indicate ingredients that appear in other MISE categories.

Chapter Nine: In the Kitchen with the Influencers

In the previous two chapters, I have told the story of our Influencer Ingredients. Hopefully, this allowed you to get familiar with the science and keep things straight without too much brain cramp. Now that we're in the kitchen, you don't need to separate your omegas from nuts, seeds, animals, and the B vitamins. You can mix everything up. As you saw in the previous kitchen chapter on Master Ingredients, I talk about both strategy *and* recipes in the kitchen. Once you understand the strategy behind ingredients and how to use them, you won't need recipes because you can find or make your own. But I'll provide a handful to help you out a bit! Take a quick peek at the Influencer Ingredient list at the end of the last chapter to remind yourself of the food we're working with.

Ramping Up Your B Vitamins, the Fastest and Easiest Ways

Let's keep it really simple and work with our nuts and seeds first. There is nothing faster than grabbing nuts and eating them as snacks, sprinkling them on salads (or any dish as a matter of fact), or eating them pulverized in the form of smoothies, salad dressings, or sauces like pestos. You don't need a recipe to simply throw nuts and seeds into anything or onto anything. How easy is that?

Now, if you want to enhance the nutty flavor or nuts and seeds and give them a little more crunch, you'll want to toast them. You can Do This in a skillet or an oven or toaster oven. Skillet toasting takes a bit more care and attention, so I use the oven method. Heat your oven to 350°F. Put your nuts or seeds on a small sheet pan. Toast them for ten minutes or until they have taken on a darker hue. Smaller nuts and seeds color up faster so take a peek at about seven minutes and remove, or continue to cook. Stay close to the oven so you can pull your jewels out at the right time. If you overcook them (they are truly too dark), toss them out and start again.

One note on buying and storing nuts and seeds: they are a luscious source of protein and fat. Their fat content makes them an easy target for oxidation or spoilage. That being said, when you can, buy them (and your herbs and spices) from a bulk bin and not pre-packaged. Bulk bins ensure turnover, meaning that ingredients get regularly dispensed and don't sit there. This reduces the potential for spoilage. Before you dispense your goodies, have a taste to ensure they are fresh. When you get them home, store them in your refrigerator or freezer until you need them. If in a freezer, transfer them to a freezer-proof container versus keeping them in a store bag. I have found that the store bags allow oxygen to penetrate, spoiling nuts and seeds or, at minimum, giving them off-flavors.

Amping Up Methylation

Nuts and seeds are great sources of many nutrients, including the vital B vitamins that drive the methylation cycle. But sometimes a variant in your genes affects how those vitamins are accessed and utilized by this important cycle. This creates a backup at the entry point into the cycle - like an automobile accident jamming up access to a traffic circle. Fortunately, we have a detour around this potential traffic jam, and it comes in the form of betaine, otherwise known as trimethylglycine. We find betaine naturally in the Chenopodiaceae family of plants (often referred to as the Amaranthaceae family) in the form of beets, quinoa, spinach and Swiss chard. Beets, spinach, and quinoa are more potent sources of betaine, which is why they are in your Genomic Kitchen Ingredient Toolbox. Egg yolks are also an excellent source of choline which is converted in the body into trimethylgycine. So an easy way to amp up your methylation cycle and ensure it "turns" is to include whole eggs in your diet. I don't need to explain how to do that do I? You know – boil them, scramble them, make a quiche or omelet. You know how to Do This....

Beets are a whole other category, causing untold numbers of people to shrink in fear of this dull root landing on their plates. Beets are both sweet and earthy in one bite. For some of you, it's the earth or the "dirt" that you taste and not the sweet notes. What's a person to do when we know functional betaine is so good for us but beets don't appeal? OK, you can avoid them and choose eggs or quinoa to get your betaine sidekick. Or, another trick I have learned is to pair them with fruit, something that actually works for all root vegetables if you are a root-vegetable shirker. Head back to Chapter Six and make the Root Vegetable Slaw recipe, only make it with Golden (Orange) beets. Oh and buy beets with the leaves on. Once you have made the slaw, take 3-4 of those leaves, pile them on top of each other, roll them up and finely slice or chiffonade them. Fold them into your slaw. Why? Read the next paragraph.

Other Ways to Eat Your Beets

- Simply raw: grab a handful of beets with the leaves on. Rinse the leaves and peel the skin off the bulb. Grate the flesh of the bulb. Now here's the real secret: try to buy the *whole* beet plant with fresh-looking, unwilted leaves. The greens are as good a source of betaine as the beet bulb itself. If the beets are small and fresh, the beet leaves will not have a thick stem. If they do, strip it out. Then roll the beet leaves together and slice them thinly. Now you have salad "greens" that you can serve with grated beets and a nice acidic dressing. If you want to mix beet greens with spinach leaves, you now have a betaine, or trimethylglycine, powerhouse.

- Roast them. Super simple. Choose beets that are similar in size. Rinse them quickly. Put them in an oven-safe dish, preferably one with a glass or ceramic lid. If you don't have a lid, cover the top of your dish loosely with foil, taking care it doesn't touch the beets. Before you cover them, add ¼-cup of water. Pop them in the oven at 375°F. and roast for one hour. Beets are cooked when you prick them with a sharp knife, and the knife slides easily into the beet. Let the beets cool. Cut the top and bottom off them and slip off the skins with your hands. Then dice or slice them and toss with an easy mustard vinaigrette (recipe below) or acidic dressing of your choice. Orange citrus is lovely with beets by the way, as is a drizzle of the balsamic vinegar reduction you can buy in most good grocery stores or make yourself. (You can find the Balsamic Glaze recipe at the end of Chapter Four.)

- Alternatively, be my guest and peel and cut the raw beets into ½-inch chunks, and roast them! In this case, check your beets for doneness after 30 minutes. Your diced chunks are less dense than the whole bulb and will cook more quickly. Just be sure to toss the roasted beets with an acidic dressing before you eat them, or you'll perpetuate the "I hate beets" movement.

Beet-Apple Soup

This recipe, lightly adapted from a Food Network Healthy Eats classic, makes the point that beets and other root vegetables pair beautifully with fruit, which seems counterintuitive—until you taste it!

PREP: ABOUT 20 MINUTES AFTER ROASTING BEETS (ONE HOUR OVEN TIME).

SERVES 6.

GRAB THESE

½ pound red beets (about 3 medium)

2 tablespoons olive oil

1 leek (white and pale green parts only), sliced

2 cloves of garlic, chopped

1 Granny Smith apple, cored, peeled and diced

¼ teaspoon ground ginger

3 cups vegetable stock

2 tablespoons lemon juice

Salt and pepper

DO THIS

1. Preheat oven to 350° F.

2. Wrap beets in foil and roast until tender when pierced with fork, about 1 hour. Cool. Peel beets. Dice.

3. Heat oil in a heavy medium saucepan over medium-high heat. Add leek and garlic and cook 5-8 minutes until tender. Stir in apple, beets and ginger. Cook 1-2 minutes. Add stock.

4. Bring to boil. Reduce heat to low, cover, and simmer until vegetables are very tender, about 25 minutes. Add lemon juice.

5. Cool soup slightly. Working in batches, purée soup in blender. Season with salt and pepper and serve. Refrigerate when cool if serving cold.

About Quinoa and the Legumes

One of the simplest ways to work with grains or legumes is to use something I call the Bowl Strategy. Think of an Asian-style bowl dish for example. At the bottom is a layer of rice or maybe noodles or rice noodles. On top of that are vegetables which may be stir-fried or sautéed. In my bowl they may be raw or in the form of a slaw. If they are cooked, flavor is introduced in this layer with a flavorizer, perhaps with a tamari or sesame-oil style sauce. (As you'll see in detail in Chapter Fourteen, flavorizers form a category of additions to a dish that pull layers of food together, even if you are not following a specific recipe. (Everything from simple dressings or vinaigrettes to pestos and salsa.) On top of the vegetables is your protein. This could be tofu, chicken, or maybe shrimp or a fried egg. I then finish off my bowl with a sprinkling of toasted nuts or seeds or some other source or "nutrient boost" to layer in more nourishment. So you see here, we are creating a layered bowl of ingredients. Using the Bowl Strategy, we can use quinoa or our Influencer legumes as the foundation. We then add spinach as our vegetable layer, or a mixture of any vegetables from the Genomic Kitchen Ingredient Toolbox. Top that with a protein of your choice and a sprinkle of nuts and seeds.

On the next page, you'll see what the Bowl Strategy looks like in chart form. Note that this bowl example has a savory focus, but you can always create a sweeter version (like a breakfast bowl) by substituting fruit from the Genomic Kitchen Ingredient Toolbox for the vegetables. In this case, your flavorizer finish might be as simple as honey.

How To Build A Bowl One Layer At A Time

The question in building a bowl is: where do we get the flavor from? This, my trusty friends, is where vinaigrettes and pestos come in.

Build A Bowl

BASE INGREDIENTS	VEGETABLES	FLAVORIZER	PROTEIN	TOP IT OFF NUTRIENT BOOSTER
Grains	Choose seasonal	VINAIGRETTE, PESTO OR SIMPLY OIL & VINEGAR	Plant Based Legumes	TOASTED NUTS/SEEDS, NUTRIMENT OILS: WALNUT, HAZELNUT, FLAX, OLIVE
EXAMPLE: QUINOA, RICE	ENSURES NUTRIENT DIVERSITY		EDAMAME, TOFU, TEMPEH	FERMENTED FOOD:
Legumes				SAUERKRAUT, KIMCHI,
COOKED FROM DRIED BEANS OR USE CANNED	Try always to include one of these		Animal	AGED CHEESE, YOGURT OR OTHER FERMENT
Rice			GRILL SMALL AMOUNT OF PROTEIN AND SLICE THINLY	
	CRUCIFEROUS VEGETABLES OR ALLIUMS (RICH IN QUERCETIN)		Seafood	
	Can be cooked, raw or both!		CHOOSE WILD CAUGHT OR RESPONSIBLY RAISED. REFER TO SEAFOOD WATCH OR OTHER SEAFOOD TRACKER FOR CURRENT BEST CHOICES	
	STIR FRY			
	SLAW			
	CHOPPED SALAD			

While I could provide you with lots of recipes for both, truth be told you can Google them. If you don't want to, then try my go-to vinaigrette recipe below or head over to the Genomic Kitchen's "Dips, Dressings, and Spreads" on Pinterest. Find one that you like and make it. Or, as a treasured reader, click here to download my "Go-To Homemade Vinaigrettes and Dressings." The reason I mention this is because you can use a dressing or a pesto to flavor sautés, stir fries, steamed, roasted vegetables or a raw salad. In my Bowl Strategy, I use a pesto or a dressing (vinaigrette) and toss my vegetables or salad ingredients with it. Put this layer on top of your grain or legume, add your protein, and dust with those nuts and seeds. *Done.* But cooking does not have to be hard. Some of us really love the kitchen. Others don't. One thing we all know is that our body needs and loves nourishing food, and your job is to provide it. The Bowl Strategy provides an easy way to fulfill the task.

My Go-To Mustard Vinaigrette for All Occasions

If you do nothing else in your kitchen, learn to make this simple mustard vinaigrette. It is the most versatile basic "sauce" you can make. No skills or fancy equipment required. It goes with most everything including salads, sautés, seafood, you name it. Just do it. A whole new world will open up. Here's my basic vinaigrette. If you don't like mine, find someone else's. I don't care. Just do it. Oh and if you don't have shallots, so what? Just omit them and make this anyway!

PREP TIME: 10 MINUTES.

MAKES ABOUT ½ CUP OR ENOUGH TO DRESS 2 HEARTY SALADS.

GRAB THESE

2 tablespoons red wine or other vinegar, such as sherry

2 shallots, minced (or part of a small red onion if not shallots)

¼ teaspoon salt

1 clove garlic, finely minced

1 tablespoon of fresh Dijon mustard

⅓ cup olive oil

2 tablespoons of fresh herbs, minced (thyme, parsley, dill, or whatever)

DO THIS

1. Throw the first three ingredients in a small bowl and combine with a fork. Let them stand for 10 minutes. The salt and vinegar soften the bite of the shallots.

2. Add the garlic and Dijon to the bowl and whisk it into the vinegar mixture.

3. Then slowly add in the olive oil and whisk all the ingredients together until they are combined into one smooth sauce.

4. Have a quick taste and add a pinch more salt if needed. If your vinaigrette is too mustardy, vinegary, or oily, add a little water to thin it and remove some of the bite.

5. Still not to your liking? A pinch of sugar or a drop of maple syrup should do the trick. It's your sauce and your palate. Make it work.

6. Once you get to the taste you like, throw in the herbs and gently combine. Let it sit for a while to let the flavors meld.

7. Use it!

Here's another handy trick. (Skip ahead if you don't like the idea of eating raw eggs.) A colleague told me that he always makes his favorite vinaigrette with one raw egg yolk. "No," I said. "Really?" Growing up in Europe, we kept (and still keep) eggs on the counter. Usually they are local eggs so we know what the chickens are up to before they've laid their eggs. My colleague made the vinaigrette, and it was sublime. Tasty. Smooth. Rich. No taste of raw egg. Egg yolks add texture to the vinaigrette, but their big role is as a fabulous emulsifier. The yolk binds the ingredients closely together for a unique and luscious taste and feel in your mouth. If this sounds appealing, take a look at my ingredients and then add the egg yolk. Now you have a perfect recipe for your taste buds and your genes. There are tons of bioactives in the olive oil and herbs; a trimethylglycine methylation infusion from the egg yolk; and a dose of omega-3 fatty acids in the olive oil.

Easy Go-To Lemony Herbed Quinoa Recipe (because I knew you would ask)

This recipe is made with quinoa, but it can be made with any grain you like. Just adapt the cooking time and amount of cooking water you need to your preferred grain. Add additional ingredients such as spinach, arugula, peppers, herbs. It's simple. Go to town with it!

PREP: 30 MINUTES (+ TIME TO COOL QUINOA). COOK: 15 MINUTES.

MAKES 4 SERVINGS.

GRAB THESE

1 teaspoon sea salt

1½ cups quinoa

1 cup cooked chickpeas (or legume of your choice)

¼ cup Kalamata olives, rinsed and sliced (or 1 tablespoon or more of rinsed capers)

2 tablespoons chopped fresh mint

2 tablespoons chopped fresh parsley (or substitute basil, thyme, or herbs in season)

2 tablespoons finely sliced basil

2 tablespoons extra virgin olive oil

Zest of one lemon, about 1 tablespoon

1½ tablespoons fresh squeezed lemon juice (or more to taste)

DO THIS

1. In a medium saucepan bring 2½ cups water and 1 teaspoon of salt to a boil.

2. Meanwhile, rinse quinoa in a fine mesh strainer, swishing the quinoa around in the strainer with the water. Drain.

3. Add the quinoa to the saucepan, reduce the heat to a low simmer, cover, and cook for 12-15 minutes. Do not overcook or you will get a mush. Quinoa is done when it looks pearly and has popped open so you

can see the germ inside.

4. Drain the quinoa and then dump it into a large bowl or onto a sheet pan to cool. Gently fluff with a fork. Do NOT skip this step!

5. If using a sheet, transfer cooled quinoa to a large bowl and gently fold in all the remaining ingredients.

6. Taste and adjust seasonings.

If you want to change this up, substitute ¼-cup fresh chopped dill and half a diced English cucumber. For a little heat, sprinkle with Aleppo pepper. You *must* include this fabulous spicy pepper in your pantry. It is finer than red pepper flakes and makes your taste buds dance!

Note: To prevent cooking water from foaming and spilling over during cooking, add a drizzle of olive oil.

Getting Healthy Seafood on Your Plate

I know that seafood is not to everyone's taste. For some of you, it is a personal ethical decision. If this is you, bypass this next section, but remember that not all of us can efficiently convert the plant-based form of omega-3 fatty acids. If you recall, some plants are sources of the precursor of omega-3 fatty acids called alpha-linolenic acid (ALA). ALA must be converted or elongated into the user-friendly EPA and DHA forms of omega-3 to be utilized by the body. Seafood is probably the best source of EPA and DHA, which is why I focus on simple seafood recipes. For those of you who are open to some seafood, read on.

I know that some of us are a tad timid when it comes to eating seafood, often because we've not been successful seafood cooks, or some folks simply don't like the taste or smell of seafood. Seafood can be a bit finicky, with a big factor being how fresh the ingredients are. If we could get same-day catch or next-day to the table, most of us would be sublime seafood cooks. Fresh seafood needs nothing more than heat (oven, grill, or sauté on stovetop) with the lightest coating of olive oil, salt and pepper, and a smattering of herbs. The secret to fresh seafood is its simplicity. So many of us are working with slightly older seafood, or previously frozen seafood which *does* differ in flavor and texture. And many of us have avoided the canned seafood aisle (except for tuna), for lack of good recipes or because we are unsure of the quality. All this being the case, below are four foolproof seafood recipes for you to try. Three of them use canned seafood, which is a more consistent product in large areas of the country than fresh or previously frozen. And yes, these recipes have been tried and tested in many culinary workshops by many converts who enjoy new and simple ways to put seafood on their plates. So give these recipes a try! And, if you are looking for tips on how to buy the freshest seafood, check out this informative video from chef Becky Selengut. Her website is also chock full of great "How to" videos on all food-related matters.

No-Fuss Tuna Salad

A can of seafood means an easy lunch or dinner. Here's my formula for a tasty tuna salad that can be assembled in 15 minutes. A great on-the-go meal, or pack it up and take it with you on a bike ride, kayak trip, hiking, or wherever you go. If you are interested in how different canned tuna products compare environmentally, check the Environmental Working Group's Food Scores section and search for responsibly sourced tuna (or any other food product).

PREP TIME: 15 MINUTES.

MAKES 2 HEARTY PORTIONS.

GRAB THESE

1 6-ounce can tuna fish, drained

1 lime, juice squeezed and its zest

1 tablespoon olive oil

salt and pepper

1 can black or white beans, drained and rinsed

½ small carton cherry tomatoes, quartered (or ½ cup or more of chopped fresh tomatoes)

ADD IN ANY OF THE FOLLOWING:

½ cucumber, chopped into ½" dice

½ apple, chopped into ½" dice

½ avocado, chopped into ½" dice

½ cup cilantro, parsley, or mint

DO THIS

1. Gently combine the tuna with lime juice and zest.

2. Drizzle the olive oil over the tuna and gently combine. Taste and season with salt and pepper.

3. Gently fold in the beans and cherry tomatoes.

4. Add any or none of the recommended options.

5. Do a final taste test. Add a pinch more salt, if needed, or a spritz more lime juice to brighten.

Get Your Salsa On!

I have taught numerous culinary workshops around the country, often working side-by-side with nutrition experts, to create recipes and meal concepts that translate our nutrition advice to the table in easy ways. We look for health-supportive ingredients that are accessible, affordable, and easy to translate to the table. In doing this work, I discovered that the **Canned Food Alliance** is very creative resource for recipe ideas, using a combination of fresh and canned ingredients to put tasty ideas on the table.

The **salsa** recipes are fabulous and so easy to make. My favorites are: Peppered Blueberry Salsa, Pineapple-Jalapeño Salsa (I use a fresh jalapeño here), and Tomato-Citrus Salsa which is made with chipotle, and served with avocado if you want. Give this website a try. You won't be disappointed.

Curried Salmon Patties with Your Go-To Salsa

PREP: 20 MINUTES COOK: 15 MINUTES.

MAKES 12 SMALL PATTIES OR 6-8 LARGER PATTIES.

GRAB THESE

2 tablespoons olive oil

1 stalk celery cut into ¼" pieces

½ cup leeks or onions, minced

14-ounce can salmon, drained and flaked

1 tablespoon curry powder

1 egg, lightly beaten

1 tablespoon minced fresh parsley

3 tablespoons lemon juice

Pinch of cayenne (optional)

½ teaspoon each salt & pepper

Fresh lemon wedges (optional)

A jar of your go-to salsa

DO THIS

1. Preheat oven to 375° F. Line a baking sheet with parchment.

2. Heat olive oil over medium heat. Sauté celery and leeks (or onions) until softened, 3-4 minutes.

3. Gently fold in the salmon and the remaining ingredients until incorporated.

4. Grab approximately 2 tablespoons of the mixture, and with wet hands shape it into patties, approximately 1" thick and 2½" wide. Place on baking sheet. Repeat until all the mixture is used.

5. Bake 12-15 minutes until heated through and starting to brown on top. Do not overcook, or the patties will dry out. Alternatively, you can brown the patties in a skillet, 3-4 minutes each side. If you choose this

method, I recommend letting the patties chill for about an hour in the refrigerator to "set up" before cooking.

6. Serve with a squeeze of fresh lemon and a salsa of your choice on the side.

Lemony Sardine Salad for Naysayers!

I call this the Naysayer Salad because nothing makes people say "no" or "heck, yes" faster than an offer of sardines. Give this recipe a whirl and expect lots of sardine converts, requests for the recipe, and scores of people looking for a second helping. In fact, you might have to double or triple the recipe to meet the demand.

PREP: 25 MINUTES.

MAKES 4 SERVINGS.

GRAB THESE

4 tablespoons extra-virgin olive oil

2 teaspoons grated lemon zest

Juice of 2 lemons

2 teaspoons Dijon mustard

4 tablespoons finely chopped fresh parsley

2 scallions, white and green parts thinly sliced

4 celery stalks, finely diced

1 teaspoon capers (optional)

4 3.8-ounce cans olive-oil-packed sardines, drained

Coarse salt and freshly ground black pepper

Artisanal crackers or baguette for serving

1. In a medium-sized bowl, whisk together olive oil, lemon zest and juice, and mustard until combined.

2. Gently stir the parsley, scallion, celery and capers into the lemon vinaigrette.

3. In a separate bowl, using a fork, gently mash the sardines until they fall apart into large flakes. Do not pummel them into a mush!

4. Gently fold in the flaked sardines until they are coated with the vinaigrette. Taste and season with salt and pepper.

5. Serve with a chopped salad, lettuce, or on a nice artisanal cracker.

Never-Fails Baked Filet of Fish

This simple, baked fish-filet recipe works every time. You can substitute cod, barramundi, or halibut for the salmon. You can also experiment with different fresh herbs for a change of flavor and taste. The layer of nuts gives the fish some crunchy texture and a luscious boost of flavor.

PREP: 15 MINUTES. COOK: 15 MINUTES.

MAKES 4 6-OUNCE SERVINGS.

GRAB THESE

1½ pounds wild salmon filets (or substitute fish, as per above)

1 cup pecans or hazelnuts

1 small clove garlic

½ cup fresh flat-leaf parsley or ¼ cup parsley and ¼ cup fresh thyme

Generous pinch of salt

DO THIS

1. Preheat oven to 350° F.

2. Line baking sheet with parchment paper.

3. Wash fish and pat dry. (Or wash with lemon juice if preferred and rinse under cold water. Pat dry.)

4. Place fish on the lined baking sheet.

5. Combine nuts, garlic, parsley and salt in a small 2–cup food processor and process until you have small crumbles.

6. Loosely pile the nut crust onto the fish filets to coat, distributing evenly.

7. Bake in center of the oven for 15 minutes.

8. Remove, season with sprinkle of salt or to taste, and serve immediately.

What About the Animals?

Yes, it is true that I am providing recipes for the plants, fish, and plant-based "B's". In 23 years of working in nutrition science and education, I have found that most people who eat meat and poultry have their favorite recipes and don't need my help—with the exception of which cuts of meat to purchase. So no meat recipes from me. I will, however, leave you with this advice: quality over quantity. Choose the best animal proteins you can afford, and those whose source you know, whenever you can. More and more, local producers of meat and poultry are selling their product at farmers markets. This gives you the opportunity to talk with farmers about how they raise their animals. Talk to them. Know that you can try one cut of meat from one vendor at a time. Go to their farms. Visit the animals on Open Farm Day. Ask questions about how the animals live, die, and how they are prepared for you. Ask the hard questions. I have yet to meet a farmer who is not interested in telling their story and sharing the love and care of their animals.

Finally, remember that we have become a culture that essentially uses the prime cuts of meat from an animal while throwing away some of the more nutrient-rich parts. Save those bones and make a broth from them. The same goes for fish skin and bones, too, by the way. Around the world, many cultures prize parts of animals that we don't cherish and use them in their cuisine. Think of charcuterie and patés, for example. These food products reflect the use of organ meat and other animal parts that not only provide nourishment, but complete the circle of life. When we take a life, then we honor it by utilizing every part and not choosing the "best" and discarding the rest.

Food for thought....

Chapter Ten: Super Foods

The other day, a list of the "healthiest foods in the world" appeared on my Facebook timeline. "Not again," I said to myself, as this happens at least three or four times a year. Of course I had to take a peek—and was very happy to see that many of the foods in our Genomic Kitchen Ingredient Toolbox made their list. Although I have no idea what criteria they used to make their choices, at least we agree. I was satisfied that I had found a "healthy food" peer and then went about my day.

The next day I got a list of the "healthiest recipes in the world," and two things emerged from this new posting. First, it elicited a tame rant on my timeline about there being no such thing as one or more "healthiest recipes" *ever*. You'll have to search my timeline for that download. Second, of course, I just had to go look and see what exactly these recipes were and why they had attained such lofty heights. Forget taste, peeps, these are the "healthiest"— by God. I curiously clicked on the link and found six recipes that addressed individual body parts—namely skin, heart, joints—and then larger issues like menopause (oh really?) and digestion. Well, imagine that! Body-part nutrition, even at its finest, is a load of you-know-what. From my training in biochemistry, I know first-hand that food does not enter your body to influence one-and-only-one specific body part. When metabolized to its basic components, namely nutrients, food is multi-directional, providing information to all your biochemical cycles. It does not compartmentalize itself, and decide when to have a chat with your brain leaving your joints in limbo for a few hours. In defense of the recipes, they did have delicious ingredients, and I did not disagree with why they were included. What riled me was this idea of compartmentalization and segmentation, suggesting that specific ingredients only work on some body parts or health issues.

What Are Super Foods?

You are probably familiar with the term "Super Foods." It seems every month or so there's another scientific article or study which suggests that scientists have made a unique discovery. They have found a food (most often on the other side of the planet) that we have never heard of, and it appears to have superior nutritional qualities. Not too far back it was red wine, then vinegar, and, more recently, foods like blueberries, açai berries, coconut water, and kombucha are all over the media, all touted as the cure for what ails you.

In the Genomic Kitchen Ingredient Toolbox, I have collected a number of Super Foods. They are neither the latest "wow" ingredients, nor are they ingredients with benefits we never knew we needed. They are also not triple-threat foods—which are largely a myth. Advocates claim certain foods have capabilities that cross different nutritional categories and appear to accomplish a number of unrelated but dynamic things in the body. Genomic Kitchen's Super Foods are a very specific group of ingredients that have a large and very diverse nutrient profile. From a practical standpoint, if you have a defined food budget, you want to choose food that gives you the most nourishment and nutrient per bite, but at the same time fits with your budget, too. Super Foods are the ultimate value buy, providing you with a vast array of nutrients without busting your wallet. Don't throw out all the other food you have at home, which would be wasteful. Rather, if you want to maximize your food dollar and return on nutrient value, start thinking about adding our Super Food ingredients to your grocery list and edge a few other ingredients—particularly anything processed or with otherwise "empty" calories—out.

What Exactly is a "Diverse Nutrient Profile?"

Proteins, carbs, and fat are the foundation class of nutrients the body uses to produce energy and support its everyday function. Collectively, these are called *macro*nutrients. Vitamins, minerals, and bioactives are classified as *micro*nutrients because they are needed in much smaller quantities than their macronutrient counterparts. Don't be fooled by their "micro" classification though. Micronutrients are essential to supporting the myriad functions of the proteins our genes make, like blood pressure regulation and digestion. While you need them in smaller quantities, the word to focus on is "need." The more nutrients that we can pack into a bite, the more nutrients your body has to select from and operate with. Remember, proteins are the body's workers, but every worker needs a set of tools in order to do its job. Super Foods provide us these tools.

How We Found Our Genomic Kitchen Super Foods

The question you are probably asking is, how did you find *your* Super Foods? I was working on a project to create a series of nutrient roadmaps for clinicians. Why? When you work in nutrition education, you are inevitably asked for foods that are the best sources of calcium or vitamin D, or whatever nutrients are needed to emphasize and optimize specific biochemical pathways that support our health. Some of us are deficient in one or more of them, and we find this out through laboratory testing. For example, a marker for inflammation that can lead to heart disease is homocysteine. Elevated levels of this compound have nutritionists thinking that you might need a boost of folate or vitamin B6 or B12. If you work in nutrigenomics like I do, your first thought might be that you have an SNP—which you remember is a genetic coding error—that is affecting how you use one or more of these vitamins, and that might be the root cause of elevated homocysteine. To boost vitamins you are lacking, you need to select foods that are the best sources. The thrifty, savvy shopper wants to put their dollars into the *best* food sources and not the mediocre ones. So I created nutrient roadmaps for every vitamin, mineral, protein, fat, and fiber-rich food source your body needs. I then mapped out the best foods for each. This way, you know, nutrient by nutrient, which food to buy without having to look it up in a database.

As you can imagine, if you spend enough time digging in databases for this type of information, you start to see patterns. Each time I searched for the best food sources for each nutrient using a uniform scientific query, I noticed that certain foods emerged as an excellent source. In other words, regardless of the nutrient I was searching for, these foods often appeared as a good or excellent source. They did not appear every time, but they appeared frequently. As I saw patterns, I organized them in a folder, and at the end of the search process, I named the folder Super Foods. All of these are ingredients that have a robust vitamin and mineral profile. If you are a savvy food shopper, then you'll be making a *super* return on your nutrient investment and your dollar with these foods.

Aren't Bioactives Super Foods?

Why yes, you could say that they are. Remember that bioactives play a different role in the food-gene relationship. Unlike vitamins and minerals, bioactives have no nutritional or caloric value. They don't behave like vitamins and minerals, and they have no energy value like calories do. Bioactives are instrumental in initiating the process by which genes are turned on or off. Genes make proteins and once proteins are formed, they need cofactors, vitamins, and minerals, to help proteins do their work. Think of it as a two-step process. Turning a gene on, or preventing it from turning on, calls for bioactives. Once the protein is created, it calls for vitamins and minerals as cofactors or helpers. So bioactives are indeed "super" in that they are vital to the process by which we engage genes. But because they do not contain vitamins and minerals, we are not including them here in our Super Foods chapter. As you know, we've already included bioactives in our GK Ingredient Toolbox. We've got them covered, and you, too!

Why Aren't Super Foods on Product Nutrition Facts Labels?

Great question! On the current Nutrition Facts Label found on the back of almost every packaged food product, the following nutrition information is required (in the US): Calories per serving, total fat, saturated fat, trans fats, cholesterol, total carbohydrate, dietary fiber, total sugars, including added sugars, protein, vitamin D, calcium, iron, sodium and potassium.

The amount of vitamins A and C, which formerly appeared on the label is no longer required. As someone who spends a lot of time reading scientific literature and looking at how nutrients inform our innate biochemistry, I see common patterns that illustrate the critical nature of all nutrients, some standing out more than others perhaps. So this begged the question: why *only* these nutrients and not others?

This certainly perplexed me, and so I reached out to colleagues who are experts in the complexities of food label law to learn why. Research supporting the Nutrition Facts Label is informed by a collective body represented by the Academy of Sciences, the Institute of Medicine, and the Food and Nutrition Board. Final decisions about which nutrients are mandated on the label are reached after lengthy periods of scientific inquiry which include a commentary period open to professionals. Once that commentary period is closed, and the FDA has responded, a collective decision determines which nutrients must make a mandatory appearance on food labels. Nutrients are chosen for their "specific relationship to a chronic disease or a health-related condition, when there is public health significance...."

For each product, food manufacturers must show on their food label (Nutrition Facts Panel) the percentage of the established Daily Value for the mandated nutrient. Other nutrients can also be voluntarily or optionally declared on the Nutrition Facts Label, providing they have an established Daily Value, an established value for how much of a given nutrient you should routinely con-sume every day. So depending on how the manufacturer interprets the science and which nutrients they feel are of value, they can include additional nutrients on the label.

When it comes to omega-3 and omega-6 fatty acids, which are polyunsaturated fats that must be obtained through diet as the body cannot manufacture them, it gets interesting. First of all, the fat section of the current Nutrition Facts Label includes information about total fat, saturated fat, and trans fat. (Though you might have thought we had gotten rid of trans fats, they still linger in some foods!) Listing polyunsaturated fats on a label is not mandated, but it should be. You already know that omega-3 fatty acids are notoriously out of balance with omega-6 fatty acids which are more broadly prevalent in our diets. We have to work harder for our Omega-3s. You might also know that some of us do not efficiently convert the alpha-linolenic, or precursor form of omega-3 fatty acid to its user-friendly EPA and DHA forms (for more on this see Chapter Nine). Genomically speaking, some of us *are* compromised. You should understand by now the robust health value of both omegas; therefore understanding where to find them is useful nutrition information. Here's what the FDA said about making the omegas voluntary or mandatory on the food label: "We decided that, because of the lack of well-established evidence for a role of [...] polyunsaturated fatty acids in chronic disease risk and the lack of a quantitative intake recommendation, the declarations of n-3 and n-6 polyunsaturated fatty acids are not necessary to assist consumers to maintain healthy dietary practices." I profoundly disagree. So would most health experts. You can follow progress and challenges related to nutrient content and food labeling claims as they pertain to omega-3 fatty acids and other nutrients at this FDA website. Then search for the nutrient you are interested in.

Enter a Paradox

The institutions that provide scientific oversight and guidance for the Nutrition Facts Label, are not the same as the institution that provides daily food guidance for Americans. In fact, the Dietary Guidelines for Americans reflect the collective body of research from the USDA, Food and Nutrition Service, and the Center for Nutrition Policy and Promotion. In creating general nutrition guidance for the public, researchers perform rigorous scientific inquiry, which includes analyzing the food intake of hundreds of thousands of Americans. (You can review this data by searching under NHANES.) Analysis of this data informs our national dietary guidance. Here is where things get "juicy" if you'll pardon the pun. The 2015 Dietary Guidelines revealed that Americans continue to fall short in their daily intake of fiber, an ongoing issue we have not yet solved. Americans also fall short in their intake of potassium, magnesium, calcium, vitamins A, D, E, C, vitamin B12, folate, and choline, a micronutrient we find in egg yolks, for example. I am not going to bore you with the biochemical significance of all of these nutrients, but isn't it interesting that population-based data indicates a shortfall of several nutrients, but only calcium, sodium, iron, potassium and vitamin D are mandated on the Nutrition Facts Label?

Nutrient deficiencies equate to chronic illnesses because the body requires *very specific nutrients* to function. Although there are established general ranges, exactly how much we need of each nutrient varies from one person to the next. Recommended daily intakes (RDI) for you is not the same for me. Among nutrition experts, it is a *known fact* that we are chronically deficient, or unbalanced in omega-3 fatty acids (see Chapter Seven on Influencers), and also in selenium. Selenium is essential for the production of glutathione, without which we cannot detoxify many compounds that are otherwise injurious to our health. And then, there is vitamin K2 which we find in fermented foods. OK, I guess I can give the label a pass there because many fermented foods are homemade and come without a label. (You will learn more about vitamin K2 as well as ferments in Chapter Twelve, Enablers.) All this to say that national guidance in the form of dietary guidelines and Nutrition Facts Labels mirrors a collective, but rather narrow, interpretation of science and thus has limitations. The question then becomes, what should you do?

Know that your body requires a *wide variety* of vitamins and minerals, sometimes called cofactors, to support a fine-tuned system. This equates to eating widely and diversely to ensure delivery of many nutrients that will support the biochemical processes required to drive *your* (hopefully) fine-tuned human engine. By this, I mean one that is in balance, or homeostasis. A Nutrition Facts Label provides us a *tiny* bit of information about nutrients, but in reality, it's more useful to glance over the label but include in your grocery cart—and thus your pantry—an abundance of foods that provide a solid nutrient insurance policy. By adding Super Foods to the other foods in our GK Ingredient Toolbox, you will ensure your body gets the broad range of basic nutrients that it needs to go about its daily business.

Don't Blame the Soil for Our Decline in Health

Though it used to be thought that the decline in nutrient quality in our food was due to a concomitant decline in soil quality, further research has shown that, with the exception of selenium and iodine, our food has *not* seen significant nutrient decline over the past century. Research published in the Journal of Food Composition and Analysis, in 2017, provides an in depth review of the research and methodology challenges that lead to the nutrient-decline conversation. It is a fascinating read for those of you interested in this much discussed area of nutrition. Depending on the area of the country you live in, the soil composition contributing to locally grown foods may vary from other areas; so, yes, your North Dakota beetroot may be nutritionally different from the South Carolina equivalent. Additionally home garden-nurtured soil, and farming techniques that deeply enrich the soil, will produce nutritionally robust food which nutritionally depleted soil will not. A plant's nutrient value reflects what it is grown in, just like your body thrives depending on which nutrients you feed it.

The reality is that while nutritionally rich soil contributes to a robust diet, our eating habits and personal tastes are more complicit in determining health outcomes and nutrient decline than soil. This is clearly reflected in a reduction in the *variety* of foods and ingredients that appear on our plates as depicted by USDA data. Take a look for yourself. The first chart which appears in the report, **Dietary Guidelines for Americans 2015-2020 Eighth Edition,** looks at how the US population over the age of one is eating compared to government-posted dietary guidance.

Dietary Intakes Compared to Recommendations

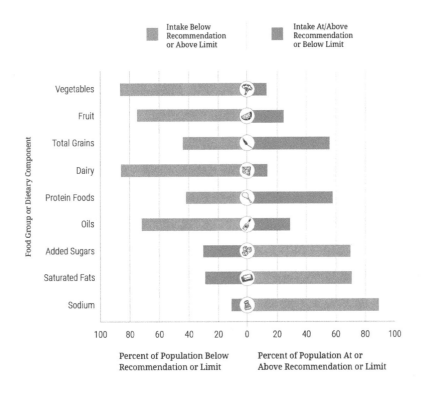

Source: US Department of Agriculture

Here is how the report summarized our status in each food category:

- Vegetables: 87% have intakes below the goal.
- Fruit: 75% have intakes below the goal.
- Total Grains: 44% have intakes below the goal.
- Dairy: 86% have intakes below the goal.
- Protein Foods: 42% have intakes below the goal.
- Oils: 72% have intakes below the goal.
- Added sugars: 70% have intakes *above* the limit.
- Saturated fats: 71% have intakes *above* the limit.
- Sodium: 89% have intakes *above* the limit.iimit.

Even though we *know* we need to eat a variety of food from different food groups, we still struggle nationally to get enough fruit, vegetables, and dairy (for those who consume them), on our plates. Now that you know how important vegetables are, particularly the cruciferous ones (many of which are dark green) take a look at the consumption data for dark green vegetables. Do you think we are getting a healthy national dose of sulforaphane? Take a look at the chart on the opposite page which compares the average weekly consumption of these vegetable jewels with recommended weekly intake among males and females over the age of one. Do I need to explain this further? For the record, these vegetables (and herbs) are included in the consumption data for leafy green vegetables: broccoli, spinach, leafy salad greens (including romaine lettuce), collards, bok choy, kale, turnip greens, mustard greens, green herbs (parsley, cilantro).

Daily Intake of Leafy Greens

■ Recommended Weekly Intake Ranges   ⬤ Average Intake

Dark Green Vegetables

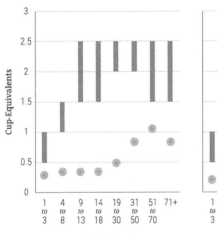

Males (years)

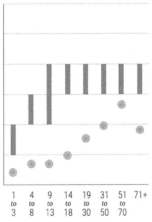

Females (years)

Source: US Department of Agriculture

Daily Intake of Fruits

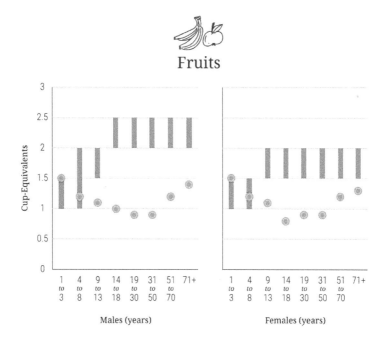

Fruits

Source: US Department of Agriculture

For every age group and across sexes, the average consumption is below the recommendation in *every* case!

Now let's turn our attention to fruit. Look at the chart above. Similar to the vegetable chart, blue represents recommended intake and orange represents actual intake. With the exception of the one-to-three age group, *all* age groups are failing to eat the fruit they should. Digging a little deeper, this is what the report said about fruit consumption: "About one-third of the intake of fruits in the U.S. population comes from fruit juice, and the remaining two-thirds from whole fruits (which includes cut up, cooked, canned, frozen, and dried fruits)." Even though our youngest population is meeting fruit intake guidelines, about 47 percent of the fruit they do eat is consumed as juice and not whole fruit. While juice can deliver nutrients, it is also a carbohydrate juggernaut, meaning it delivers one huge whack of glucose for your insulin to deal with. Moreover, do you know what happens to excess carbs? Yep, they end up as fat if the body cannot utilize their energy. Need I say more? Here's my advice. Bypass the juice aisle completely and head to the produce area. Buy the whole fruit. If you want to drink juice, press the whole fruit and drink that juice. Chances are you'll drink 2 to 4 ounces maximum. That I can live with, and your body will thank you too!

Now on the good news side, while our produce consumption is woefully low, research by the Produce For Better Health Foundation shows that we are broadening our palates, allowing more room on our plates for berries, for example. In the vegetable category, carrots, spinach, avocado, and kale are on the upswing. Potatoes remain the number one most consumed vegetable, followed by lettuce, and then onions. Broccoli consumption comes in at a distant eleventh place. Here's the thing to think about. Cast your eyes back to our Master Ingredients and you might recall that crucifers are an essential source of the bioactive, sulforaphane. Onions are essential source of the bioactive, quercetin, as are radishes (also a crucifer). The only three representatives of these two bioactives on our nationally preferred vegetable intakes are kale, onions and broccoli. And yet the cruciferous family is vast! We have *a lot* of work yet to do to insure we start eating to help -our genes function optimally. Now you know why I created the Genomic Kitchen, here are some no fuss quick tips to get those important crucifers on your plate.

How Do I *Know* which Nutrients I Need as an Individual?

Let's look at this from two angles: a traditional nutrition science perspective and a genomic perspective, using the topic of inflammation. Traditional nutrition science recognizes that we are all unique individuals with unique nutrition needs. Recommended Daily Intake guidance (RDI), now known as the Daily Values (DV), was created to provide general guidance on the average amount of each individual nutrient we need on a daily basis, depending on our age. For example, if you are over four years of age, pregnant or breastfeeding, the RDI suggests 1,300 mg calcium per day. If you have diabetes or heart disease, your nutrient needs may be different from someone who does not. In an optimal world, we would use nutrient test information from lab work to identify your personal precise needs. This information indicates your nutrient needs at a given point in time, which obviously can change over a lifetime, and depends on whether you are an athlete, a pregnant mom, or a person dealing with a health issue, for example. So that's the traditional nutritional approach. Now let's consider the genomic angle.

You recall that gene SNPs, or Single Nucleotide Polymorphisms, are mistakes or variants on one or more genes within a biological pathway that *might* impact how the body absorbs or utilizes a nutrient. SNPs, when found, are a signal to check whether nutrient availability, or the body's ability to use a given nutrient, might be compromised. And many times your nutrient levels are just fine. Genomic information provided by personalized genomic testing is *one* of several tools that we can use to evaluate your nutrition status and provide precise food guidance. Truth be told, most of us will do just fine if we serve ourselves, as we have learned, from our steadily growing Genomic Kitchen Ingredient Toolbox. These ingredients provide a solid nutrient foundation for your body to operate from, and can provide a consistent, well-rounded nutrient infusion, the likes of which your body may never have experienced!

A Nutrition Facts "Label" for Oxidative Stress and Consequent Inflammation

Let's use oxidative stress to showcase how we might construct a more informative Nutrition Facts Label that better relates to the body's food-gene relationship. Oxidative stress, or a bombardment of free radicals, causes inflammation if we don't sequester or "extinguish" those radicals. Oxidative stress, left untamed, leads to inflammation. To tackle it, we can certainly increase the number of antioxidants we eat from our diet because they can mediate those pesky free radicals. The antioxidants we get from food do not, however, work as effectively as the ones we can prod our genes to produce. We call the gene-produced antioxidants, *endogenous* antioxidants. So for example, we can eat more foods rich in vitamins A, C, and E to increase the number of antioxidants. (And where are they now on the food label?) In addition, we can selectively choose foods that stimulate the powerful firehose system via Nrf2 that we talked about in Master Ingredients, Chapter Five.

In the form of an equation, an optimal intake looks like this:

Bioactives + Nrf2 *plus* protein supporters (vitamins and minerals) = successfully contained fire = reduced oxidative stress.

Bioactives + Nrf2 *minus* protein supporters (vitamins and minerals) = uncontained fire=oxidative stress causing inflammation.

The bioactives we need to activate Nrf2 are lycopene, curcumin, quercetin and sulforaphane. The nutrients we need to support the proteins (enzymes) produced by the genes that Nrf2 activates are copper, iron, manganese, selenium, and zinc. Add in the antioxidants you get from your diet—vitamins A, C and E—and you have a nutrition roadmap to tamp down on oxidative stress. So, look below to see my Nutrition Facts Panel. If this were my food product, I might also include helpful language, such as "this food product contains nutrients that help your body manage oxidative stress." If you are wondering why I put omega-3 fats on my panel but didn't discuss them above, you know that these fats can turn off Nf-kB, thereby reducing inflammation that can cause oxidative stress. More exciting, recent science shows that EPA and DHA are omega-3 fats that directly act on reactive oxygen species in the cell, as well as activate Nrf2 in vascular cells. This activity suggests yet another cardio-protective role for this jewel of a fat, which confers protection by mitigating oxidative stress!

MY NUTRITION FACTS PANEL:

- Fats: Omega-3
- Bioactives: sulforaphane, curcumin, lycopene, quercetin
- Vitamins: A, C, E
- Minerals: Selenium, Manganese, Copper, Zinc, Iron

You see, when you use nutrigenomics + biochemistry, it helps you navigate beyond food claims to ingredients whose behavior you understand.

Back to Super Foods

As you may have noticed from my "equations" above, foods with a robust nutrient profile, namely those with lots of vitamins and minerals per mouthful, are essential to our biochemistry equation. In the Genomic Kitchen, we purposely set out to find and create a shortlist of these foods to help you efficiently support the business of your body's biochemistry. Super Foods are *purposely* included in our GK Ingredient Toolbox to provide so many of the vitamins and minerals you need. Keep them on hand in your kitchen to accompany the selection of ingredients from each of the MISE categories, and you will have the nutrient foundation you need to support the biochemistry that *is* you.

Announcing the Super Foods

If you haven't turned the pages to peek at these foods yet, here they are:

MY SEVEN SUPER FOODS

1. Chickpeas (garbanzo beans)

2. Flaxseed

3. Lentils (red, black, green--all kinds)

4. Sea vegetables (wakami, dulse, kombu, kelp)

5. Sesame seeds

6. Whole soybeans

7. Sunflower seeds

Remember, I chose these foods because their robust nutrient profile kept appearing each time I searched for different nutrients such as magnesium, zinc, or calcium. Here are three things to remember.

1. Each of these foods is not the best source of *every* nutrient! (That is marketing hype.) They contain a wide variety of many nutrients, but do not claim to be the best source of *all* individual nutrients. Think of them as excellent all-rounders, so to speak.

2. If we put all of these foods in a basket and analyzed their collective nutrient profile, some nutrients might still be under-represented, such as omega-3 fatty acids. Super Foods are selected to act as a nutrient insurance policy for your diet, ensuring an influx of many vitamins and minerals to shore up your biochemistry and support your health. Use them alongside ingredient choices from each MISE category to maximize their benefit.

3. This list is not an exclusive list. Super Foods appear in this category because of the search criteria I used. Don't throw out all the food you currently have on hand; just start adding these to your pantry.

Now that you understand everything you need to know about my Super Foods, it's time to head over to the kitchen and dig deeper into each of them, and, more importantly, get them onto your plate.

Come with me....

SUPER FOOD INGREDIENTS

Legumes	Seeds	Other
Chickpeas	Flaxseed*	Sea Vegetables*
Lentils	Sesame Seed*	
Soybeans*	Sunflower Seed*	

Asterisks indicate ingredients that appear in other MISE categories.

Chapter Eleven: In the Kitchen with Super Foods

I am going to keep this Super Foods kitchen chapter nice and easy for you to navigate. I'll unpack each ingredient and show you simple ways to get each one on your plate. No gourmet cooking skills required!

Chickpeas (Garbanzo Beans)

With a robust spectrum of nutrients, chickpeas—like lentils—truly deliver! They pack a complete punch by providing all essential and non-essential amino acids (non-essential denoting that the body can make them), which are the building blocks for protein. Chickpeas are used in their whole and ground form (as flour or meal) in many countries. You'll find them in curries in India, soups and salads in Europe and America. They're also the main ingredient of hummus and smashed down and rolled into delicious falafel in the Middle East. (Hummus means generally using them with a tahini sauce—made from sesame seeds, a super food sibling!—so yeah!) Different cultures also have their version of flatbreads made from chickpea flour. The Italians call their version *farinata*, the French *socca,* and the Indians, *pappadum.*

Easy Things to Do with Chickpeas

- Use as base for dips and spreads.
- Make into patties and falafel (then freeze).
- Add to soups and grain-based salads.
- Roast them for a snack.
- Throw them on top of a chopped salad.
- Eat them cooked or rinsed right out of the can.
- Use them as a base for a curry.
- Use to dredge or thicken as a substitute for wheat flour.

Make an Easy-peasy Pasta and Chickpea Soup

Here's a soup that I adapted from a book that focused on quick after-work recipes. The chickpea-tomato purée provides a nice, thickened base, resulting in a simple hearty soup. No fuss. No muss. Use the lardons, or omit if you want. Cheese is also optional.

PREP: 10 MINUTES COOK: 25 MINUTES.

SERVES 4.

GRAB THESE

I tablespoon olive oil (plus more to drizzle)

2 ounces pork lardons (optional but adds nice base flavor!)

I medium white onion, chopped in ½" dice

2 medium cloves garlic, minced

3 large juicy tomatoes, rough chopped

1 tablespoon tomato paste

1 tablespoon fresh rosemary, rough chopped, or 1 teaspoon dried

1 15-ounce can chickpeas, drained and rinsed

2½ cups unsalted chicken or vegetarian stock, store-bought or homemade

5 ounces dried pasta (macaroni, fusilli, orecchietti or farfalle for example)

1 tablespoon fresh thyme, finely minced or 1 teaspoon, dried

Salt and pepper to taste

1 ounce grated Gruyère, Emmentaler, or parmesan cheese

DO THIS

1. Heat a medium saucepan over medium heat and add the olive oil.

2. Add the lardons (optional) and sauté a couple of minutes to release their juices and aroma.

3. Add the onions and reduce the heat until you just hear them sizzle. Cook about 4 minutes, stirring occasionally, until they are soft and you can smell that delicious aroma.

4. Add the garlic and heat, stirring gently for a minute.

5. Meanwhile, using a small food processor, purée the tomatoes, tomato paste, rosemary, half the chickpeas and 1 cup of stock (use water if you prefer)

6. Stir the puréed ingredients into your onion mixture in the pan, along with the remaining stock. Add your pasta, bring to a simmer, cover, and cook according to directions or until *al dente*. Add more stock or water for a more liquid soup or to prevent the ingredients from sticking, if necessary.

7. Stir in your fresh thyme and have a quick taste, adding salt and pepper to your preference.

8. Gently stir in the remaining chickpeas and heat through for 1 minute.

9. Serve in bowls with a whirl of olive oil and a sprinkle of cheese (optional).

To add more to the food-gene cross influence, throw in a handful of arugula or finely sliced kale just before serving. Now you've got yourself some grab-and-go sulforaphane too!

My "Go-To" Tabouli Salad for Pot Lucks, Boat Trips, Easy Supper, and Beyond!

Tabouli salad is the easiest salad to put together, making you look like a rock star chef, though it requires minimal culinary aptitude or effort. While traditionally made with toasted cracked wheat (bulgur) and chickpeas plus parsley, you can switch this up however you want. Use quinoa and lentils if it pleases you. Switch out the parsley for mint or basil, or both. Of course, if you do so, it is not technically a traditional tabouli, but you get the point. Recipes provide proportion and directions for assembling the dish correctly.

PREP: 10 MINUTES COOK: 35 MINUTES.

SERVES 4.

GRAB THESE

1 cup cracked wheat (bulgur wheat)

2 cups water

1 cup grape tomatoes, or summer tomatoes rough cut into ½" bites

3 scallions, white and green parts sliced into ½" pieces

4 ounces feta cheese, roughly crumbled or cubed

½ cup Italian parsley, minced

Zest of 1 lemon

2 tablespoons fresh lemon juice (more if preferred)

2 ounces olive oil

Sea salt, cracked black pepper to taste

¼ cup sliced Kalamata olives (optional)

DO THIS

1. Pour the bulgur wheat in a bowl and cover with two cups of boiling water. Cover with a lid or plate and leave until the bulgur has absorbed

all the water, about 30 minutes. Cool and then fluff up with a fork.

2. Gently fold in the tomatoes, scallions, feta cheese, parsley, and lemon zest.

3. Pour over the lemon juice and oil and then light stir until evenly distributed.

4. Taste and add salt and black pepper to your preference. Add olives if you are using them. Enjoy!

Note: You should know by now that the above salad is calling out for a handful of spinach, watercress, or arugula. Or maybe a sprinkle of radishes or some other embellishment from the GK Ingredient Toolbox!

Go-To Chickpea Dip with a Curry Twist

When we think of chickpeas and dip, they are often paired with tahini to make a hummus. Now, tahini is itself a Super Food because it is essentially a sesame paste and sesame seeds are one of our Super Foods. So hummus is a go-to super recipe! Use a tablespoon or so of tahini in this chickpea dip recipe, or make it as is. You see, when you know how food works in your body, you can adapt any recipe to substitute the ingredients you prefer!

PREP: 10 MINUTES.

MAKES 2 CUPS.

GRAB THESE

¼ cup raisins or golden currants (sultanas)

1 15-ounce can chickpeas, drained and rinsed

Zest of ½ lemon

2 tablespoons fresh lemon juice

1 tablespoon olive oil

1 teaspoon curry powder

1 teaspoon ground ginger

2 tablespoons water (or more as needed)

½ teaspoon kosher salt

DO THIS

1. Plump up raisins/currants in a little warm water. Set aside.

2. Whirl all remaining ingredients in processor until smooth.

3. Stir in raisins or currants just prior to serving. Adjust salt.

Want more food-gene zip in your dip? You know what to do! Whirl in a handful of greens. Who cares if your chickpea dip turns green?

Lentils

In the US, you'll mostly find four types of lentils: red, green, brown, and black. Arguably, there are two colors of green if you ask the experts! There's the brownish green and then the deeper "French green" or Puy lentils, but that's splitting hairs. Lentils are not only a *robust* source of vitamins and minerals, but also of protein and fiber. The perfect all-rounder! Lentils are also nutritionally blessed with some of the bioactives that positively influence our genes and create better health, including apeginin, luteolin, myricetin, quercetin, and coumaric acid. Score!

Easy Things to Do with Lentils

- Make lentil soup (and add curry for a one-two punch).

- Cook up a *dal*, which is essentially curried lentils (recipe below) to go over your brown rice.

- Precook green or brown lentils and throw them in a lentil salad in the place of other grains.

- Use them as a base for dips and spreads in the place of black beans or garbanzos.

- Add cooked, mashed lentils to ground meats before cooking to provide an alternative plant-based protein source and extend the volume of the dish.

Lentil Essentials

Not all lentils should be treated the same. This means that you don't cook the orange ones the same way you cook the others, for example. They are smaller and cook faster. One cup of dried lentils makes about 2½ cups of cooked. In general, use 2½ cups of water to one cup of lentils. I use orange lentils for soups, although I admit I love them in *dals*. Use the heartier brown and green lentils for salads and yes, *dals,* because they hold their shape better after cooking. Use any form of lentils in patties, spreads, dips, and purées.

Simple Red Lentil *Dal*

Growing up in the UK, I ate Indian dishes from an early age and have continued eating them all my life. I love the pungency and myriad of flavors combined with the heat. Every Indian family has their own version of traditional recipes and how they mix spices to create those memorable flavors. This red lentil *dal,* which follows the basics of a foolproof recipe by chef John Gregory Smith, has been reproduced by hundreds of folks in my workshops. All of them become enthusiastic converts to the simple, yet powerful flavors. I adapted the recipe to include greens, and the use of canned tomatoes when the fresh ones don't deliver on juice or flavor, which is sadly all too often.

PREP: 10 MINUTES COOK: 30 MINUTES.

SERVES 4.

GRAB THESE

1 cup red lentils, washed and drained

2 tablespoons olive oil or high oleic sunflower oil

2 teaspoons cumin seeds

1 red onion, finely minced

1 inch ginger root, grated or minced (or 1 teaspoon dried ginger)

1 green chili, seeded, finely chopped

3 large in-season juicy tomatoes, chopped into ½" pieces (or 16 ounces canned diced tomatoes)

½ cup water (more as needed)

1 teaspoon turmeric

2 teaspoons ground coriander

1 teaspoons sea salt

2 cups baby spinach, or fresh spinach torn into 1" pieces

Chopped cilantro for garnish

DO THIS

1. Bring 2 cups of water to a boil in a medium pot.

2. Add the lentils and a dash of olive oil to reduce foaming. Bring back to a boil and then reduce and simmer for 10 minutes. Cook until the lentils are soft and cooked through, but not reduced to a mush. There will still be some water in the pot. Drain the lentils and set aside.

3. Remove the wok or skillet from the heat and gently fold in the tomatoes.

4. Add in the remaining spices and salt and cook gently over a low heat for 8 to 10 minutes, stirring occasionally.

5. Fold reserved lentils gently into the mixture, then fold in the spinach and heat for a couple of minutes until it is wilted.

6. Remove from the heat, garnish with cilantro, and eat as is, or serve with rice.

TIP: If the canned tomatoes do not give enough richness of flavor to the *dal*, stir in 1 tablespoon of good quality tomato paste. This is why choosing quality ingredients, or working with fresh produce in season is what we should all strive for! Juicy fresh tomatoes should break down into a thick sauce, but do add in water as needed to ensure a sauce forms. If using canned tomatoes, you should have enough juice to make a saucy mixture.

Soybeans

Ah yes, the much maligned soybean. It has been exalted and denigrated in recent times. Much of the denigration results from its role as an oil crop, which detracts from its rich nutritional value. Yes, it is important to acknowledge that some of you may not tolerate soy in some form or any form. Obviously if this is you, you can make other Super Food choices and get the vitamins and minerals your body deserves. That being said, if we focus the conversation away from soybean oil and onto its whole, unprocessed "bean" form, we once again have a robust, easy food product to work with in the kitchen.

Easy Things to Do with Soybeans

- Buy them frozen in the pod. Follow package instructions for cooking. Sprinkle with a good quality salt, or seasoning of your choice. Peel away and pop in your mouth. Delicious!

- Use canned (rinsed and drained) or frozen, shelled soybeans as a legume in any salad or dish where legumes are called for.

- Use soybeans as a "sprinkle" on salads, meaning, throw a handful or two, or three into a salad and toss.

- Soybeans, when dried, become soy "nuts." You can buy them as a snack, or make them yourself. Making them yourself requires starting with the dried beans and soaking them overnight. You then dry them off before tossing with oil and spices and roasting. If this is for you, just get out there on the web or Pinterest and find a recipe that speaks to you.

Minty Avocado Edamame Spread

With this ten-minute recipe, you can swap out your regular mayo for something more luscious, or just use it as a dip or a dollop (I'm thinking with your scrambled eggs, tofu, chicken, etc.). Edamame are soybeans that are harvested earlier on in the harvest season when they are less mature, and therefore softer. In this recipe, feel free to switch up the herbs too, substituting the mint with basil, cilantro or a parsley-cilantro mix.

PREP: 10 MINUTES.

MAKES 2 CUPS.

GRAB THESE

1 cup frozen shelled edamame, cooked per package instructions and cooled

2 avocados, halved, peeled and pitted

1 teaspoon minced fresh ginger

1 clove minced garlic

¼ cup water

3 tablespoons fresh lime juice

1 teaspoon fresh lime zest

¼ teaspoon sea salt

½ to 1 teaspoon wasabi powder (optional)

2 tablespoons fresh chopped mint

DO THIS

1. Place all ingredients in a small food processor and process until all are incorporated and smooth.

2. Taste and adjust seasonings. Add a pinch more salt or wasabi to heighten flavor. Add more lime juice or zest to add brightness.

Flaxseed

Flaxseed is cool in so many ways. Nutritionally robust, it is an excellent source of alpha-linolenic acid (ALA) which most of us can convert to the user-friendly form of omega-3 fatty acids EPA and DHA! Flaxseed is also an excellent source of a class of polyphenols (those non-nutrient compounds) called *lignans*. Lignans are transformed by intestinal bacteria into enterolactone, which has been well studied. Its key health attributes appear to be:

- Acting as antioxidants to prevent DNA damage;

- Preventing oxidation of lipids which is associated with cancer;

- Managing hypercholesterolemia, hyperglycemia, and atherosclerosis; and

- Interfering with elevated levels of endogenous estrogen activity associated with both breast and prostate cancer. This means occupying estrogen receptor sites which might otherwise be occupied by stronger endogenous estrogen that the body produces.

Sesame seeds, also on our Super Foods list, are another excellent source of these valuable polyphenols.

Flaxseed contains mucilages which help your gut do what it does best; namely extract the nutrients it needs and then get rid of the stuff it doesn't need. Foods with a high mucilage content are also useful in the kitchen as a replacement for eggs in baked products, such as a waffle or pancake batter.

Here's a tip!

If you mix three tablespoons of water with one tablespoon of flaxmeal and let it sit for about ten minutes, you'll have a gel-like product similar to the texture of a raw egg. You can then use this to bind together other ingredients. I sometimes use it instead of an egg when I am making curried salmon patties, for example. So don't try to fry your flax slurry like an egg, but use it as wet ingredient when your recipe calls for an egg to bind dry ingredients!

Easy Things to Do with Flaxseed

- Add it to anything you put in a blender or food processor: soups, smoothies, dips.
- Throw it on anything you might add nuts or seeds to: salads, breakfast cereal, oatmeal.
- If you bake, use it as a raw ingredient (whole flaxseed, or ground), even added to pancake and waffle batter.
- Use as an egg substitute.
- Use it as a thickener.
- Roll slices of banana in it and add to a plate of fruit for some texture and flavor contrast.

Sesame Seeds

Sesame seeds are among the oldest oilseed crops and happen to have some of the highest oil content of all seeds. Examples of other oil seeds are sunflower, soybeans, rapeseed (canola), and flax (also known as linseed). You may be familiar with hulled sesame seeds used to adorn bread and buns, but across the world the seed takes many different forms. In the Middle East, sesame is ground into tahini, a paste that you might be familiar with if you eat authentic hummus or *halvah*, that sweet confection-like product with the texture of flaky fudge and the taste of honey. In Japan, sesame seeds are mixed with salt—and sometimes seaweed—to make *gomashio*, a versatile seasoning consumed in many ways.

Easy Things to Do with Sesame Seeds

· Sprinkle on breakfast cereals, salads and just about in or on anything.

· Add to smoothies.

· Add to homemade energy bars or balls.

· Add to bean-based patties and falafel.

· Throw sesame seeds into the mix or batter when you bake.

· Make your own *gomashio* condiment (see Sea Vegetables at the end of this chapter).

· Grind up fresh sesame seeds to make a basic tahini paste and use it in hummus and salad dressings. Or just buy a jar and keep it on hand in the refrigerator.

No-Bake Sesame Apricot Balls or Bars

PREP: 30 MINUTES.

MAKES AS MANY OR FEW BALLS AS YOU WANT, DEPENDING ON THE SIZE.

GRAB THESE

¾ cup almonds

1¼ cups unsulfured apricots

2 tablespoons nut or seed butter

2 tablespoons sesame seeds

2 tablespoons melted coconut oil

¼ teaspoon salt

1½ teaspoons ginger (or more to taste)

Pinch cayenne (optional)

DO THIS

1. Line a jelly roll pan or small tray with parchment paper or wax paper.

2. Pulse the almonds in a food processor to a coarse meal the texture of breadcrumbs.

3. Add all remaining ingredients and pulse until combined.

4. Dig into the mixture with a tablespoon and grab a healthy spoonful.

5. Using wet hands (it's easier, trust me), slide the mixture off the spoon, into your hands and roll into a sphere the size of a golf ball. Place the ball on your lined tray and repeat.

6. If you hate the idea of food on your hands, use an ice cream scoop or make bars by dumping the almond mixture into a small, lined square pan (say 8" x 8") and pressing it flat. If you want fat bars, don't press to hard. If you want thin bars, press harder.

7. Refrigerate or freeze.

For balls, I recommend refrigerating the mixture for an hour, uncut. For bars, just freeze the entire pan! Then, cut into bars. They are now ready to eat, or wrap each ball or bar individually to snack on later.

Sunflower Seeds

Selenium is one of my nutrient darlings and yet it also the one most deficient in Americans' diets. But, good news! Sunflower seeds are a great source of this baby! Selenium is one of the power nutrients we need to support the enzymes that combat oxidative stress and enable detoxification. This food is truly a veritable powerhouse capable of tackling many tasks at one time.

One of the most beautiful sights in North America is sunflowers growing on our Northern Plains against the backdrop of blue sky on a late summer day. Sunflower seeds are the fruit of the sunflower. Most sunflower seeds are grown for their oil and are solid black. The striped sunflower seeds, however, are the ones we consume as an ingredient. Who knew? While we grow a lot of sunflowers here in the US, Russia, the Ukraine and the European Union actually grow more! Just like Americans chew gum, our Russian friends chew sunflower seeds.

Easy Things to Do with Sunflower Seeds

- Make sunflower seed butter (to use instead of a nut butter).
- Add them to pestos and dips, substituting other nuts or seeds.
- Toast or roast them for snacks and add to trail mix.
- Use as toppings or additions to salad.
- Add to granola, granola bars, energy bars.
- Add whole or ground into veggie burgers.
- Grind up and use as a flour to dredge.

Now Do This!

Make a pesto. Ha! I am not going to tell you how. Go to Pinterest, and grab a recipe. Easy. Okay, okay! If you want an even easier route, go to my Pinterest boards. (All you need is an email address to use the site.) Go to the Super Foods board and scroll. You'll find a recipe for sunflower seed pesto, I promise you. Now, if you get distracted along the way, you'll find lots of other recipes there too. If you don't find one using sunflowers, then make something else with another Super Food.

Sea Vegetables (or Seaweed)

Say what? Don't recoil in shock. Yes, I really am referring to the stuff that washes up on the sea shore. Not that I am recommending you just pick that up and start dumping it in your cooking pot. I prefer you seek out seaweeds (sold at many larger groceries and health food stores) that have been selectively harvested by companies who do this as a business. They understand the ocean, the growing environment, and how both can work to provide you with a quality product. (That being said, in 2017, I did harvest and eat seaweed washed up on the pristine shores of the west Ireland coast. Simply memorable and also delicious!)

What do I mean by sea vegetables? Specifically, look for wakami, nori, kelp, kombu, dulse, spirulina —ever heard of these? (Though most are dried, some restaurants are starting to serve some seaweeds fresh, like sea beans and some kelp.) They are usually off in the international—usually Asian—section of your grocery store, though more and more brands are being harvested in coastal states like Maine and Oregon. They don't come with a "Super Food" label or a flashing "Buy me! Buy me!" light, but in my opinion, they should. I describe sea vegetables as "one of nature's most perfect foods that most of us have never eaten, but should." The nutrient makeup of sea vegetables varies by region, type, and indeed season, so my recommendation is to buy different types of sea vegetables, and vary your choice among companies to take advantage of their changing nutrient profiles. After all, sea vegetables are like plants, they reflect the environment they grow in.

Beyond the flavor they impart, one of the *many* things I love about sea vegetables is their natural iodine content. Yes, indeed, seaweed is your go-to iodine delivery system. No need to buy commercial iodized salt any more. Just add some sea vegetables to your salt grinder and you make iodized salt with a big nutrient punch. Make this your *new go-to condiment*! Oh, and if iron is on your nutrition needs list, look no further than your friendly sea vegetables, too.

Easy Ways to Use Sea Vegetables in the Kitchen

- Use it like a condiment! Sprinkle on, or into smoothies, dips, pestos, stir fries.

- Crumble or flake into soups, onto salads.

- Add a sheet of kombu to your pot of legumes or grains as they cook. Seaweed imparts flavor and, once consumed, also helps dissipate gas, generated by some vegetables' glutamate content, for those of you who have a tough time with delicious legumes. Here's another trick. Once you've used it, fish it out, let it dry on a clean kitchen towel, then use it again! Recycled food, what a novelty!

- Make fun and delicious homemade sushi (with seafood or vegetarian ingredients).

Sesame Seed and Seaweed Gomashio

PREP: 5 MINUTES. COOK: 12 MINUTES.

MAKES ABOUT 1 CUP.

GRAB THESE

½ cup unhulled sesame seeds

½ cup sea vegetables (dulse flakes, or crumbled kelp or nori, or a bit of each!)

1 teaspoon coarse sea salt (the large crystal type)

DO THIS

1. Heat your oven to 350°F.

2. Spread the sesame seeds out on a small baking tray, place in the oven and let them toast until a light even brown, about 12 minutes. Remove from the oven and cool.

3. Mix together the toasted sesame seeds, sea vegetable(s) and sea salt in a small bowl.

4. Pour into your salt grinder or shaker and use! (You may have to lightly pulverize in food processor or mortar and pestle first.)

And here you are! You made it to the end of Super Foods. I hope can see how flexible and easy they are to add to your plate. You don't need to get fancy with recipes because these ingredients are so easy to nip and tuck into dishes. Remember that Super Foods are super because they deliver a hearty portion of the vitamins and minerals (cofactors) your proteins need to do their job. By including them as often as you can, you ensure your body receives the right nutrient information to get its business done!

Chapter Twelve: Enabler Ingredients—a Key to Your Gut

Many of you may have heard the expression, "You are what you eat." In reality, "You are what you absorb," or what you *can* absorb. If you cannot absorb the nutrients and the bioactives from the food you eat, your genes don't have a whole lot to work with. Think of the gut—comprised of the small intestine stretching many feet from stomach to colon—like a garden. The health of the garden is determined by the quality of the soil, which is determined by how robust its nutrient composition is. Healthy soil is the foundation for healthy plants. In the same way, your gut requires a healthy foundation to function optimally for a strong body that stays in balance. Just as poor soil yields sickly plants, so do poor nutrition, stress, and environmental toxins lead to a "sickly gut" which will contribute to chronic health issues.

The gut is the key to the body's nutritional foundation. If it is not breaking down and distributing food properly, then your body will not be able to receive the nourishment it needs to function optimally. A gut out of balance effects your body in many ways, including food sensitivities and intolerances, outright allergies, auto-immune diseases, mood swings, and mental anxiety and depression, to mention just a few issues.

Enabler Ingredients optimize the health and function of the gut so that you get the nutrition you need from the foods you eat, and the health you deserve. Let's take a deeper dive into how the gut works. Then we'll look at how the probiotic and prebiotic foods—which make up the Enablers—work synergistically to enrich our health.

Gut Science from the Inside Out

Think of the gut wall like a fortress, allowing only those nutrients that have biological permission to pass from the intestines into the body. Imagine windows with heavy bars firmly cemented into the frame. Bars in these windows prevent escape. Over time, the bars may become loose as the cement they are set in is exposed to the elements. Eventually, the cement crumbles, the bars fall out, and our prisoner escapes!

The gut wall separates the outside world (the food you have ingested passing from mouth to anus) from the inside world where nutrients are absorbed and put to work. Cells lining the lumen, or inside wall of the gut, adhere tightly to each other. This tight adhesion prevents the passage of any-and-all things that your body deems foreign or toxic from escaping from the intestines and into the body. One of the many roles Enabler Ingredients play is to help fortify the gut wall by providing the right bacteria. In turn, these bacteria provide nutrients and fuel the cells that make up the lumen and line the gut. More on those bacteria a little later.

The Root of Digestive Problems: Inflammation, Stress, and Gut Function

As we know from previous chapters, inflammation lies at the root of many chronic diseases. A body out of balance will fire up the immune system, and the first line of much of your immune defense relies on the health of your gut. You will see why as you read through this chapter and learn that your gut is like the retaining wall of a fortress. It should not be penetrated. Many things can interfere with the normal functioning of that biological fortress wall, including stress, poor nourishment, exposure to environmental toxins, consuming foods that do not agree with us, and trauma, which naturally weakens our defenses by taxing our immune system. (Some resources: general discussions of digestive function and gut health and the neurobiology of stress.)

How do we know if our gut has become dysfunctional? Perhaps the easiest way (and I am being very simplistic here) is to observe the speed at which food transits through the digestive tract. You know this as either diarrhea, regular bowel movements, or constipation. In medical terms, we call this gastrointestinal motility. Physiological changes in the body can have a dire impact on how well your gut works, and this manifests itself in a chain reaction of events like a series of dominoes falling. First one, then in a cascade until there are no more left standing.

As food makes its way through the gut, a progressive process of simplification occurs, reducing food from a complex mass into its simplest nutrient parts. These simple parts become the nutrient information the body can "read," translated as "incorporate and work with." The right amount of saliva and chewing in the mouth starts the food breakdown process. The correct level of acidity in the stomach continues the nutrient "decoupling" process which continues into the small intestine and then the large intestine. This complex series of transformations enables food in its right (digestible) form to be presented at the gut wall for absorption and to be used in the body's cells. Given that this is a journey, you can see how suboptimal levels of acidity in the stomach prevent an effective process of breakdown, resulting in poor absorption of nutrients farther down the tract. Or worse still, if our gut wall (fortress) is compromised, fragments in food that are either toxic or not in a form the body recognizes are allowed to be absorbed, causing an immune reaction. Strange molecules are then "read" by the body as threats and, therefore, need to be annihilated. An immune reaction can manifest itself in a number of different ways, some sneaky and some "in your face." Sneaky means you might start to react to foods you previously always tolerated, but you can't figure out which food and why. Overt means that you essentially feel ill. It may start with an occasional headache and become a migraine, maybe fatigue, insomnia, listlessness, a general feeling of malaise, skin rash, and, all too often, a mix of symptoms that appear to be unrelated, but truly are. If you are "fortunate," your body recognizes intruders before the invasion occurs and food is rapidly excreted as diarrhea, preventing internal harm, but also causing a loss of otherwise useful nutrients.

Other times, ineffective digestion can slow down the passage of food and the opposite reaction occurs, constipation. All this to say that your gut is connected from your mouth all the way to your anus, and what happens between these entry and exit points deeply impacts whether you absorb nutrients, lose them, or if that food simply festers in your gut.

Stress and the Gut

We may have all experienced that feeling when stress makes us run to the nearest bathroom. No fun. The question is what connects stress to the gut and why? In short, the lining of the gut, otherwise known as the intestinal mucosa, is woven through with nerve fibers that are influenced by neurotransmitters emitted by the brain. In other words, the brain communicates directly with the gut, and vice versa. When the body senses stress, or you are a "bundle of nerves" before an interview or an exam, the brain prompts the body to release Corticotropin-releasing Factor (CRF). Release of this hormone results in a "direct order" to your colon; the end result of which can be a sudden release of all of its contents. Just lovely! Research with animals injected with CRF demonstrates the connection between this powerful hormone and diarrhea. In other words, stress talks to your gut. For more on the powerful connection between stress, gut, and signals from the brain, you might want to browse this landmark book, *The Second Brain* by Dr. Michael Gershon. Or, take a peek at these articles from Harvard Health, and this article from Johns Hopkins.

Equally important is the impact of chronic stress on your gut. In my early days as a clinician, I would often see young women serving in the Armed Forces coming to me with digestive complaints. Their frequent complaints were abdominal pain, which was sometimes labeled as "phantom pain" by doctors. Abdominal scans and gastrointestinal testing usually did not reveal the root cause of their pain and discomfort, and yet it was very real for them. Research in human social genomics, a field of study that analyzes the impact of the environment, socio-economic status, isolation and social threats, sheds light on this pain. Specific to dysbiosis (the state of having a poorly or dysfunctional gut), a rich body of scientific literature supports the connection between its manifestation, often in cases of Irritable Bowel Syndrome (IBS) and childhood trauma and abuse. The latter could be take the form of parental neglect, physical abuse, sexual abuse, life-threatening events such as fire or an earthquake, violence or disruption of the family (divorce). I am certainly not saying that all the women who came to see me had exposure to the aforementioned, but I *am* saying that stress can be very subtle and have a long lead time before the physical effects show themselves.

Simply put, stress in all its forms has a long-lasting and deleterious effect on our health. Its entry point is frequently emotional, which, in time, impacts the health of the gut. Chronic stress is destructive to human nature and human health. Regardless of how nutritious our diet is and how carefully our food is grown, a beautifully constructed diet will never overcome the negative impact of chronic stress on our health. This means that tending to the environment we live in and the relationships we have (or don't) is as important as the food we put in our mouth. Food can only "fix" so much.

Dysbiosis

In recent years, there has been a huge uptick in the number of people suffering from a condition called *dysbiosis*. This means that the gut is not functioning properly or is unbalanced. When the gut is not in working order, it creates turmoil in the body. That turmoil can manifest itself in symptoms as simple as headaches, aches, pains and fatigue, or as devastating as auto-immune diseases like Irritable Bowel Syndrome (IBS) or Crohn's Disease. These diseases can make you feel really sick but not necessarily know how or why.

Dysbiosis happens when your gut and its microflora get out of balance; it is a reflection of this state. Our essential gut bacteria (or microflora), which we'll be looking at in more detail shortly, have a number of different roles. Gut wall protection is among the most important. Gut bacteria in the colon ferment indigestible plant fibers to produce short chain fatty acids (SCFA). One of the SCFAs, butyrate, is a principle fuel source for the cells, or enterocytes, that line the gut wall and are integral to shoring it up. Without healthy enterocytes, the tight adhesions between the cells break down. When that happens, foreign substances that should be expelled from your body start to find their way out and into places they should not. Your body's response to foreign substances is a battle cry: your immune defenses start to rally, and rally, and rally. In fact, they don't stop heeding the battle cry until the gut wall is secured—and it can take a *long* time for your immune system to put away its guns. The biochemical guns go back in the armory *only when* the gut wall is secured again.

The Essential Role of Healthy Bacteria in Our Gut

Your gut maintains a critical balance of beneficial bacteria which play a number of important roles which we'll be discussing shortly. The figure on the next page highlights the many roles of the microflora, or bacteria, in your gut. On the left hand side, you see that they play an important role in manufacturing a variety of crucial nutrients, and improving their bioavailability. On the right hand side, you can see their importance as a first line of defense for the body, neutralizing harmful substances like environmental toxins and helping the body combat potential threats from disease.

When the Microflora Get Compromised

Studies show that, starting at birth, babies delivered via the birth canal have a more robust microflora than those who are not. Natural vaginal delivery allows microflora from the mother to pass to the infant (a process called colonization) and begin the process of building the infant's immune system. We all start life with a very primitive immune system, passed from mother to infant. The mother's immune system is the starting blueprint for the infant's own immune system, and microflora are the immune information delivery system. Without vaginal delivery, contact with the mother's microflora is absent. Scientists believe this is *one* reason why children born via cesarean section may have higher rates of asthma or celiac disease versus children born via the birth canal.

The health of your digestive tract from the stomach down effects your microflora. Consider pH, or relative acidity, which

The Power of Microflora in The Human Gut

HOW THEY HELP WITH NUTRITION	HOW THEY SUPPORT YOUR IMMUNE SYSTEM	ROLE OF SPECIFIC BACTERIA IN IMMUNE SYSTEM
Improve Bio-Availability Of Nutrients	Provide Fuel To Enterocytes That Line The Gut Wall	Bifidobacteria
	HEALTHY ENTEROCYTES MAKE A STRONG GUT WALL WHICH PREVENTS ENTRY OF PATHOGENIC BACTERIA	THE MOST POPULOUS IN THE GUT. PREVENT COLONIZATION OF VIRULENT MICRO-ORGANISMS
Manufacture B-Complex Vitamins And Vitamin K		
		Lactobacilli
Ferment Indigestible Carbohydrates (Prebiotics) To Produce Short Chain Fatty Acids (SCFAs)	Produce Anti-Viral Sub-stances	PRODUCES LACTASE AND HYDROGEN PEROXIDE WHICH ACT AS POWERFUL ANTISEPTIC, ANTI-VIRAL AND ANTI-BACTERIAL AGENTS. MUTE CYTOKINE RESPONSE
	Produce Anti-Fungal Sub-stances	
SCFA: ACETATE, PROPIONATE, BUTYRATE		
	Neutralize Endotoxins	
	TOXIC SUBSTANCES PRODUCED BY SOME GRAM NEGATIVE (PATHOGENIC) BACTERIA	
	Regulate Cytokines	
	CYTOKINES ARE SIGNALING COMPOUNDS INVOLVED IN COORDINATING IMMUNE RESPONSE	
	Neutralize Carcinogens	

must be within certain levels for the gut to function optimally. As I mentioned earlier in this chapter, when stomach acid is low, food can pass into the digestive tract at the wrong pH or acidity level. This sets up a cascade of events that not only impact the availability of enzymes to digest your food, but may actually change the pH of your entire gut. Yes, I am talking about both the small and large intestine. Changes in the pH of the stomach have a downstream impact on the sheer quantity and composition of bacteria in the small intestine. Research has shown a correlation between changes in the stomach pH and a fairly common condition called SIBO, or Small Intestinal Bowel Overgrowth. In the large bowel, or colon, changes in the pH render the environment less favorable to your friendly, beneficial bacteria, and more favorable to pathogenic bacteria that do not serve you at all—and may harm you. Widely prescribed medications known as Protein Pump Inhibitors appear to play a key role in changing the pH of our intestinal tract and subsequently the critical makeup of our microflora.

Other compromising factors include medications, particularly antibiotics, as well as injury, stress, surgery, environmental toxins, and age itself. What you eat can have the biggest impact of all. Remember, food is information. If we provide the wrong kind of information through the food choices we make, then we fail to provide the microflora of your gut with the information they need to help us feel our best.

The Gut-Weight Connection

One area of research where we know there is a strong connection is between dysbiosis and the real life struggle many of us have with our weight. Many people complain of daily efforts to maintain a healthy weight, which may have nothing to do with dietary habits. Advances in science investigating the composition of gut bacteria reveal that an imbalance between the major bacterial players (known as *phyla*) in the gut correlate with obesity or weight struggles. Simply put, there is a relationship between your bacteria and your weight and that relationship refers to a delicate balance between the *firmicutes* and *bacteroidetes* bacterial strains, which dominate in the gut. In a much publicized study appearing in *Nature* in 2006, researchers reported that the higher the proportion (or ratio) of *firmicutes* to *bacteroidetes* in the gut, the greater the propensity to obesity. Researchers also observed that when they translocated the *firmicutes* strain of bacteria into lean mice and lean human volunteers, changing the normal ratio of *firmicutes* to *bacteroidetes* they observed a significant change in total body fat. The reason for this is that *firmicutes* have an uncanny and greedy habit of extracting more energy (calories) from food passing through the gut than their *bacteroidetes* counterparts. So guess what? More calories translates into more fat storage which means you gain weight. Manipulating this delicate balance in the gut through food and probiotic supplementation is one of many tools clinicians can use to help individuals attain a healthy weight.

The Satiety Connection

There are myriad connections between what we eat now and how it impacts our health with one critical area being the brain-gut connection. As you've learned, a major function of the bacteria in our gut is to ferment specific fiber components in the food we eat. The process of fermenting these fibers produces compounds called Short Chain Fatty Acids (SCFA). These specialty fatty acids come in many forms, but most important to the gut are propionate, butyrate, and acetate. One of the major differences between a diet deficient in plant fibers and one replete with them is the amount of measurable propionate, butyrate, and acetate compounds present. These three compounds are significant for a number of reasons.

Butyrate is a compound synthesized by the *firmicutes* strain of bacteria (yes, we need *firmicutes,* even if the bacteria is a culprit in weight gain when not in balance with *bacteroidetes*). And, as you have already learned in this chapter, butyrate is the primary fuel used by the cells that line the walls of the gut. No butyrate means a compromised gut fortress, which means a whole host of health problems for you.

For the weight conscious among us, the *acetate* component of the SCFAs produced by the bacteroidetes strain is a unique tool in our nutrition tool chest. Research shows that acetate can activate a communication or "signaling" process via the cells lining the gut. Activation of these signals communicates with hormones on the other side of the gut that themselves communicate with the brain. There are many outcomes to these exchanges of signals. From a weight management perspective, two of the hormones activated are GLP-1 and PYY, glucagon-like peptide-1 and peptide tyrosine respectively. These two hormones have the uncanny ability to shut down your appetite, or turn off your hunger mechanism. Now think about this. Your body is ingenious. If you seed your gut with food containing the right kind of bacteria, and then you feed that bacteria with the right kind of plant-based foods with the fiber these bacteria crave, the bacteria will, in turn, produce compounds that automate your weight-control mechanism. In other words, bacteria can be instrumental to how we manage satiety.

And that's not all! The same *acetate*, *propionate*, and *butyrate* compounds also positively affect your emotional wellbeing, too. In the same way that they communicate with hormones that work with your satiety mechanisms, they communicate with hormones that influence your mood. How about that! Now each of us is different, and, remember, food is only one tool in your health toolbox, albeit a very important one. But it is impressive to think that by seeding your gut with the right bacteria, and then fueling them with the right kinds of food, you can influence the way you feel, and have another great day.

Shore Up Your Gut Defenses: Enter The Enablers

The Enabler Ingredients include both probiotic-rich fermented foods *and* prebiotic foods that are rich in specific fibers that gut bacteria utilize. Let's explore both of these so that you may better understand their role in supporting your gut. I'll start with fermented foods.

The terms "fermented" and "cultured" are used interchangeably when talking about food, and refer to the process of reducing a whole substance into a less complex substance, with the help of bacteria, yeasts, or fungi. It is an anaerobic process (does not use oxygen) in which microorganisms convert carbohydrates, for example, into energy (ATP) and by-products. Fermentation also generates carbon-dioxide as a by-product giving fermented beverages like kombucha (and beer!) their effervescence. The process of fermentation produces a variety of different bacteria which we find in fermented foods from pickles and miso to yogurt and kimchi. The word to note here is *bacteria*. The fermented foods we are interested in are rich in gut-friendly bacteria and are referred to as *probiotic-rich* fermented foods. (Read this to learn more about fermented foods and probiotics in overall diet.) It is the fermented food with living (probiotic) bacteria that confer the health benefits I have been discussing. This is important, because many foods and beverages are created using the *fermentation process*, but the end product does not contain the living bacteria we seek. Sourdough bread uses a living yeast-starter culture that provides a characteristic taste; however, baking kills the yeast in the culture. Wine and beer are also fermented products, but the yeast used in the fermentation process is most often largely removed from the final product.

Definitions aside, what I personally love about these products is that they were born of practicality. Each of them is traditionally a way of both preserving the harvest, but also extending the life of the food itself. It just so happens that they are also some of the best natural "medicine" we can ingest—and you can produce them yourself with no Nutrition Facts Label required.

Fermented Foods—Beyond Yogurt and Sauerkraut!

Fermented foods are cultural foods as the chart below shows. Different global cultures create their own favorites and integrate them in culinary preparations to enhance flavor, and, perhaps unknowingly, nutrition. Yogurt is perhaps the most globally recognized food, but geography and the dominant kind of animal milk at

Examples of Traditional Fermented Foods

Dairy
YOGURT
GREEK/BULGARIAN-STYLE YOUGURT, YOGURT, FILMJOLK SKYR

BUTTER
CULTURED BUTTER, BUTTERMILK

MILK/CREAMS
CLABBERED MILK, SOUR CREAM

KEFIR

CHEESE
FARMERS CHEESE, AGED CHEESE

Dairy-free
NUT MILK YOGURT, NUT MILK KEFIR

Fruit
PRESERVED LEMONS, LACTO-FREMENTED FRUIT CHUTNEYS, FRUIT BUTTERS, JAMS

Vegetables
SAUERKRAUT, SAUERRUBEN, PICKLED GREENS, SOUR PICKLES, SWEET PICKLES, PICKLED GARLIC, PICKLED BEETS, ETC.

Beverages & Tonics
BEET KVASS, HERBAL TONICS WITH PROBIOTICS, KOMBUCHA, REJUVELAC, WILD CIDER AND MEAD, NATURAL SODAS, WATER KEFIR

Grains
BREAD
SOUR DOUGH

Legumes
MISO, NATTO, TEMPEH, SOY, TAMARI

Protein
MEAT/FISH
SALT-CURED, DIRED (JERKY)

hand impart significant differences in flavor and texture. The strains of bacteria used to culture the milk, be they naturally occurring or physically introduced, creates the differentiators. Icelandic-style yogurt differs quite significantly from Bulgarian-style yogurt, for example. Cultures introduced in cheesemaking highlight the difference between Roquefort, cheddar, brie, and other cheeses. People around the world ferment a wide variety of grains using different cultures, resulting in fermented foods with different dietary applications. In Burkina Faso and Ghana, for example, millet is fermented to produce *ben-saalga* liquid or "gruel," often used for weaning infants. *Idli* is a savory rice cake produced by fermenting rice and lentils and is a popular breakfast food in parts of India, Malaysia, and Sri Lanka. You are likely familiar with soybeans fermented with specific mold strains to produce foods as different as *natto*, miso, and *shoyu* in Japan. Some cultures ferment fish or shellfish to produce fish sauce condiments which star in culinary preparations throughout Asia. In other countries, notably Spain, Italy, India, Thailand and Turkey, fermented meat products are frequent additions to the diet in the form of sausages. Regardless of their origin, all of these products are an inclusive part of the diet and plate, and unknowingly provide nutritional advantage to the consumer.

Why Fermented Foods?

As I mentioned, the process of culturing or fermenting essentially takes raw ingredients and breaks them down into simpler, edible parts, generally with that easily identifiable tart or even effervescent taste. Take cabbage. When you ferment it, you retain the crunch, but the cabbage takes on that familiar acidic or "pickled" flavor. This is the result of lacto-fermentation, or the work of *leuconostoc* and *lactobacillus* bacteria. Vinegar is also a fermented product, the work of both yeast and bacteria. Yeasts first change the natural sugars in the base food to alcohol through "alcohol fermentation." Acetobacters then convert the alcohol into vinegar in acid or acetic fermentation. The result is an acidic product that we use robustly in cooking to balance flavors or to make them shine.

The process of fermentation creates a nourishing and useful product that is easier to digest and which contains more nutrients, notably zinc, B vitamins—folate, riboflavin, vitamin B12, vitamin K 2—and some amino acids. But equally important, fermented foods increase the number of beneficial bacteria which we know as probiotics in the gut. These bacteria help tilt the balance in your gut away from less advantageous or sometimes pathogenic bacteria. Remember gut imbalance and weight gain? Fermented foods are one food-based tool we can introduce into the diet to help correct these imbalances, or prevent them altogether. Our ancestors were clever, even if they did not know that they were making their best medicine simply by preserving the harvest.

So, fermented foods allow us to 'seed' the gut with beneficial bacteria (also called probiotics), produced by the fermentation process. By "seeding" we mean infusing the gut with beneficial bacteria, which can balance out harmful, pathogenic bacteria.

Foods containing probiotic bacteria are also referred to as probiotic-rich foods. You may be familiar with *lactobacillus* and *bifidobacteria,* which are produced during fermentation. You can find these good bacteria in various forms on grocery and health food store shelves as supplements or added to yogurt.

Depending on the food being fermented, different strains of bacteria will be produced, each of which plays an important role in the gut. Fermented dairy products may be an excellent source of vitamin B12, folate, and biotin, all produced by the lactic-acid bacteria in the fermented product. Fermented vegetables include sauerkraut and kimchi and pickles. Depending on the food, fermentation may increase levels of iron or zinc. It may also reduce phytic acid, which binds up nutrients, rendering them inaccessible for absorption. Fermented foods may not only increase the number of beneficial microflora in your gut, but they are also a valuable source of nutrients themselves and enable better absorption. You could say that fermented foods are nutritionally superior foods. Take a look at the chart on the next page to see just how nutritionally valuable they are.

Fermenting Versus Pickling: What's the difference?

You might be wondering what the difference is between pickling and fermenting.

Fermentation and pickling are both a common form of food preservation. They both create that characteristic acidic or vinegary flavor. The difference is that the natural process of fermentation does not use heat or the addition of vinegar to preserve the food. Naturally fermented food is also a probiotic, literally a living food. Many commercially produced pickles you buy in the store use vinegar and heat in the production process to preserve the food, extend shelf life, and to kill potential food-borne pathogens. Vinegar and heat essentially kill the fermentation process, making these

Health Value of Fermented Foods

SYNTHESIS OF NUTRIENTS	RICH IN ENZYMES AND BIOACTIVES	STRENGTHEN IMMUNE SYSTEM	IMPROVES DYSBIOSIS	INCREASES FOOD LIFESPAN	IMPROVES DIGESTIBILITY AND REDUCES ANTI-NUTRIENTS
Vitamins FOLATE, RIBOFLAVIN, B12, K2 SYNTHESIZED BY SPECIFIC STRAINS OF BACTERIA	Heat applications can destroy natural food enzymes and bioactives NATURALLY FERMENTED FOODS RETAIN FOOD ENZYMES	Beneficial bacteria deters pathogenic bacteria in the gut	Fermented foods re-populate the gut with beneficial bacteria	Fermenting foods is a form of natural food preservation DAIRY: AGED CHEESE, KEFIR VEGETABLES: KRAUTS & PICKLES FRUIT: CHUTNEY & PICKLES	Reduces phytic acid (a natural plant defense mechanism, which can bind and prevent absorption of some nutrients) SOAKING & FERMENTING FOODS REDUCES THIS BARRIER INCREASED ABSORPTION: IRON, CALCIUM, ZINC
Amino Acids		Produce Short Chain Fatty Acids PROVIDE IMMUNE DEFENSE FUEL TO SOURCE FOR CELLS THAT LINE THE GUT			
Short Chain Fatty Acids					
Secondary Polyphenols BY-PRODUCT OF DIETARY POLYPHENOLS					
Other COMPOUNDS WITH NEUROTRANSMITTER FUNCTIONS (GABA) LACTATE REDUCES ROS IN INTESTINAL CELLS					

pickled products void of all bacterial life. Commercial pickling does not produce an environment in which natural yeasts or bacteria transform raw food into a fermented product. Unless the product says lacto-fermented or cultured, "pickling" is simply a generic term that indicates the food has been bathed in vinegar with added herbs and spices. It does *not* necessarily indicate that probiotics are present.

Fermented Foods In Action: Secret Weapons for Heart and Bone Health

By now, I hope you are really starting to connect the dots between the food you eat and the power of your gut. For decades, researchers have been looking at the health implications of fermented

N THE GENOMIC KITCHEN

foods—a mainstay in diets around the world. Whether they are in the form of sourdough starters, miso, kombucha, aged cheeses, yogurts, or more, they have significant beneficial implications for our health. Personally, I feel that they are the best food-based "medicine" we can put on our plate on a daily basis. Here are two more reasons for you to put them on your grocery list—or better still, make them yourself.

It is no secret that, on average, the leading cause of death worldwide is heart disease. Depending on where you live, this statistic may or may not apply due to occurences of infectious disease or death related to exposure to contaminated food or unclean water. Regardless, if we tease apart the data, heart disease dominates the globe, followed closely by cancer. Osteoporosis is also a modern disease that is creeping up the ranks in terms of prevalence, too. The reason I mention this is because fermented foods can be a huge health assist with all of them. Let's start first with bones.

Many of us have been taught that healthy bones require daily intake of calcium and vitamin D. While both of these nutrients play a critical role in forming the bone matrix required for bone strength and health, they are just two players among a larger cast. Also, research has shown that another nutrient has quietly entered center stage. That nutrient is vitamin K2, a member of the vitamin K family. You may be familiar with vitamin K1 (phylloquinone) and its role in coagulation or blood clotting. But the K family has other members, and a star for bone health is vitamin K2; confusingly also known as both menaquinone and MK-7 and found naturally in fermented foods.

Osteocalcin is a protein hormone that is critical to the bone-building process. It works by taking calcium from the blood and binding it to the bone matrix. In order to carry out its bone building duty, osteocalcin must be in the right (active) form—and vitamin K2 is the osteocalcin activator. In other words, if you want to ensure the health of your bones, you need to add fermented foods to your diet, because they are the natural source of this critical nutrient.

Research supports these basic facts: vitamin K2 activates osteocalcin through a process of carboxylation, or fixing calcium to the bone matrix. Researchers have found that a high rate of unactivated (or uncarboxylated) osteocalcin in the blood equates to a six-fold increase in hip fractures. So it's not just about getting the calcium into the body, it's about how we get calcium (and other nutrients) to bind to the bone matrix itself.

Now a very rich fermented soy food source of vitamin K2, called *natto*, is a product routinely eaten in Japan. It is definitely an acquired taste, and its texture and pungent flavor are not likely to be found on the average dining plate in America. That being said, compared to other fermented products, it is extremely rich in vitamin K2 and also an excellent source of calcium and magnesium, both instrumental to bone density. When researchers performed bone mineral density (BMD) tests on the hip sites in 944 women ages 20 to 79, they noted that habitual natto consumption correlated with higher bone mineral density numbers among post-menopausal women. This is significant because as women age their levels of estrogen diminish—hence the estrogen replacement therapy often prescribed for post-menopausal women. The relationship between estrogen and bone health is too complex to describe in detail here, and indeed would distract us from our focus on fermented foods. Suffice to say, that estrogen plays a critical role in offsetting or blocking the inflammatory process that can cause bone breakdown or "resorption" if left unmanaged. For the scientifically curious among you, dive in deeper into the role of estrogen and bone metabolism here.

Looking again at natto, in a landmark study researchers

compared the incidence of hip fracture among women living in eastern and western regions of Japan. The dietary intake in the eastern part of the country is characterized by high percentage of vegetables, as well as natto. The opposite is true in the western part of the country. The difference in hip fracture rates between the regions was quite startling. Women in the eastern regions had significantly lower rates of hip fractures than the west. Controlling for all variables, researchers attributed the differences in hip fracture rates to vitamin K2 the eastern women consumed in natto.

And if you need more reasons for putting fermented foods on your plate, here it is: vitamin K2's cousin, vitamin K1, which helps with blood clotting. Many of us know that leafy greens are an excellent source of Vitamin K1, and yet many of us run in the other direction at the grocery store. Here's a little secret for you. You *need* that vitamin K1. Even though you might be busily stacking fermented foods for precious K2 on your plate, if the body senses a deficit in this important nutrient, then vitamin K2 will be converted to K1. So yes, K1 will help clot your blood, but now you don't have enough vitamin K2 available to stick calcium to your bones! While this is a simplistic explanation, it does serve to illustrate the importance of a wide and diverse diet. Our forefathers were smart. They grew many of their vegetables and ate them right out of the garden. They also knew how to preserve and extend the harvest by culturing foods, preserving dairy in the form of yogurt and cheese, and pickling extra vegetables the natural way, creating lacto-fermented sauerkrauts, pickles and condiments. In short, they got their K1 and their K2, too!

Now, About Fermented Foods and Your Heart

We've seen that vitamin K2 is a critical element in insuring the strength of your bones, but K2's contribution does not stop there. When we are talking about heart health, calcium once again is at center stage. K2 directs calcium to your bones, but what happens when this pivotal vitamin is not around—when you aren't eating your fermented foods? Research has established that unattended/unbound calcium has a tendency to wander into your arteries and accumulate there. Scientists have known for a long time that excess calcium, when deposited into the arteries, ends up as part of the plaques which are so injurious to the vessels themselves. Injured blood vessels lead to something called endothelial dysfunction, or vessels whose lining is compromised and are therefore incapable of doing their job properly. When our blood vessels are compromised, blood does not move around the body efficiently. Your heart ends up working harder to do the job. In the end, an overworked heart is an ineffective and weakened heart. Your body compromises by working harder, which leads to higher blood pressure and a poorly functioning pump. How does K2 help?

Vitamin K2 is activated by a protein called the Matrix Gla protein, abbreviated as MGP. In this scenario, the active form of vitamin K2 steers calcium away from the vessel walls, in effect preventing the formation of pesky plaques. And where does K2 put that calcium? In your bones of course! The question is, how do we know K2 is effective in reducing cardiovascular disease? We turn to the science of course.

A landmark longitudinal project called the Rotterdam Study followed 4807 healthy men and women 55+ years old. Researchers observed that a high intake of K2 (approximately 32 micrograms/day), correlated with a 50-percent reduction in arterial calcification and cardiovascular risk. When they considered the role of Vitamin K2 against a wider source of risk factors (mortality from all causes), they correlated the dietary presence of the vitamin with a two-percent risk reduction. In another study, the Prospect-EPIC, studying a cohort of 16,057 healthy women aged 49 to 70, a higher intake of K2 was shown to be cardio-protective. In this case, for every ten micrograms of K2 consumed on a regular basis, the risk of heart disease was reduced by nine percent. To put this in perspective, a cup of whole fat yogurt has approximately 0.49 micrograms of K2. One ounce of hard aged cheese has 2.3 micrograms. Sauerkraut 0.2 micrograms and 100 grams (about 4 ounces) of natto a whopping 939–998 micrograms. Now, I will leave you with this point: currently there is no established RDA for Vitamin K2, even though research is showing just how important this vitamin is. In the US, the RDA for Vitamin K (which presumably means the whole family of Vitamin K) is: 90-120 micrograms. Specific to vitamin K2, depending on life stage and whether bone disease is present, current research suggests a recommended daily intake of 45-120 micrograms a day.

Do you get the point? There is a reason to eat your fermented foods.

There Are No Coincidences

In her fascinating book, *The Jungle Effect: Healthiest Diets from Around the World*, Dr. Daphne Miller recounts her experience taking a sabbatical from her busy medical practice to investigate the diets consumed by people in communities with a lower incidence of specific chronic diseases, such as depression or diabetes, around the world. It is important to note that these communities were not void of every chronic disease, but rather had lower incidences of some of these diseases. The communities she investigated were in the Copper Canyon, Mexico, Iceland, Crete, Okinawa, and the African country of Cameroon. In her work, Miller made a number of fascinating observations. The two that stood out the most to me were the consistent use of particular herbs and spices and the regular consumption of naturally fermented foods in the diets of these healthy populations.

Connect the dots with me here. We know that herbs and spices have a nutrigenomic capability, meaning they are foods that deeply influence our genes. You now also know that fermented foods are a source of beneficial bacteria that are essential to the integrity of your gut, and subsequently your health. Now these two observations alone did not completely explain why these particular communities had lower incidences of specific chronic disease. For that, you'll have to read the book to find out the whole story! But you now know from this book that herbs, spices, and fermented foods contribute greatly to your health.

A Summary of Additional Health Benefits of Fermented Foods

They contain and/or make available more vitamins—B vitamins, magnesium, zinc, and possibly Vitamin D.

The process of fermenting grains (as with sourdough starter) produces bioactive compounds that may confer immune and anti-inflammatory benefits and improve antioxidant status and glycemic status.

Probiotics (yogurt) may improve antioxidant status.

Fermented foods have a higher antioxidant capacity than their non-fermented equivalent (fermented versus unfermented soymilk for example).

The isomalto-oligosaccharide compounds that are found in traditional fermented foods such as miso and soy sauce appear to promote growth of the highly beneficial *bifidobacteria* and *lactobacilli* found in our guts.

In laboratory research, even fermented herb formulations have been shown to confer anti-inflammatory benefits in animal models.

And One More Thing

About those bioactives that you are now so familiar with: researchers have been able to demonstrate that specific bioactives (in this case quercetin, rutin, caffeic acid, and chlorogenic acid) not only stimulated the proliferation of *bifidobacteria* (one of the major beneficial gut bacteria), but also decreased the ratio of the troublesome *firmicutes* to *bacteroidetes* bacteria. In plain English, we know that when the number of *firmicutes* bacteria exceeds *bacteroidetes*, we see imbalances in the gut which can manifest as weight gain. Polyphenols also appear to stoke or stimulate bacterial production of short chain fatty acid production. And now you know the benefits of those short chain fatty acids. So, once again, we have a reason to get those bioactives on our plates.

Are Naturally Fermented Foods for Everyone?

You may have gathered by now that I am a fan of fermented foods, and the science I have described tells you why. All this being said, not everyone can tolerate them, at least in high doses. If you have never tried fermented foods, I suggest you introduce them slowly and in a few bites (or sips) per day. They are a living food, and their bacteria and effervescence may initiate changes in your gut that create some gassiness or mild discomfort. This soon shall pass as you get used to them. Some people, however, have gene variants that mean they cannot efficiently break down the histamine content of fermented foods. This intolerance can manifest in a variety of ways, including itchy skin and headaches, both mild and severe. If this is you, I recommend you discuss your symptoms with a licensed/accredited healthcare practitioner and find how and when you can incorporate these essential foods without causing distress. Usually, it is a case of small doses of them over days, versus large quantities on one day. Get the help you need.

Prebiotics

I have dedicated a lot of time to probiotics, or beneficial bacteria. Now it's time to talk about their sidekick, *prebiotics*. Without prebiotics, probiotics are ineffective, dead in the water, if you will. Beneficial (probiotic) gut bacteria require specific fibers as food, shown in the chart opposite. These fibers are fermented by

How Fiber in Specific Food Supports Your Gut and Your Health

INSOLUBLE	HYDROPHYLLIC	SOLUBLE	OLIGOSACCHARIDES
Cellulose INCREASES SHORT CHAIN FATTY ACIDS (SCFA)	**Supports growth of SCFA: Propionate**	**Specific Fiber: Oat Beta-glucan** SUPPORTS GROWTH OF SCFA.	**Inulin** SUPPORTS GROWTH OF BIFIDOBACTERIUM
FOOD SOURCE: FRUIT, VEG, LEGUMES, SEEDS, NUTS	**Supports growth of Bacteroidetes, Firmicutes & Bifidobacteria**	SUPPORTS GROWTH OF LACTOBACILLUS, BIFIDOBACTERIUM	FOUND IN CHICORY ROOT, BURDOCK ROOT, JERUSALEM ARTICHOKES
Hemicellulose INCREASES BETA GLUCURONIDASE TO SUPPORT DETOXIFICATION	FOUND IN GUMS SUCH AS ZANTHUM, GUAR ARABIC	FOUND IN OAT, BARLEY, AND RYE	**Fructo-Oligossacharides** SUPPORTS GROWTH OF SCFA
FOUND IN WHOLE GRAINS		**Specific Fiber: Mushroom Beta-glucan** SUPPORTS GROWTH OF SCFA	SUPPORTS GROWTH OF LACTOBACILLUS & BIFIDOBACTERIUM
		SUPPORTS HEALTHY BACTEROIDETES: FIRMICUTES BALANCE	FOUND IN ONION, CHICORY, GARLIC, AND ASPARAGUS
		Pectin SUPPORTS GROWTH OF SCFA: BUTYRATE	**Galacto-Oligosaccharides** SUPPORTS GROWTH OF SCFAS: ACETATE & LACTATE
		FOUND IN APPLES, PEARS, PLUMS, APRICOTS, CITRUS (PEEL), BLACKBERRIES, RASPBERRIES, STRAWBERRIES	SUPPORTS GROWTH OF LACTOBACILLUS & BIFIDOBACTERIUM
			FOUND IN LEGUMES

bacteria to produce nutrients and other compounds that support the integrity of the gut. Think of them as the food and fuel source our beneficial gut bacteria need to thrive. Collectively we call foods containing these fibers *prebiotics*. One of the terms we use in medicine to link probiotics with prebiotics is "seed and feed." Through fermented foods we *seed* the gut with beneficial bacteria which are produced during the fermentation process. Once the gut is seeded with these bacteria, we need to *feed* them with food containing the fibers they require to thrive and do their work. We call those feeding foods prebiotics. Probiotics *seed*. Prebiotics *feed*.

Foods containing insulin and compounds called fructo-oligio-saccharides are commonly referred to as prebiotics. Prebiotic-rich foods include artichokes, asparagus, bananas, chicory, garlic, leeks, jicama, onions, soybeans, wild yam, and whole wheat products. Enablers include not only fermented foods, but also a basket of very specific other foods that serve to fuel the probiotics present in fermented foods. It is a one-two punch to better nutrition and therefore overall health.

Finally, here is the master list of the Enablers.

Note: in the fermented foods section of Part Two: Enablers in the Kitchen that follows, I provide examples of popular fermented foods. There are many more for you to discover. This just gives you an idea.

So, without further ado, let's head into the kitchen and figure out how we can get some of these important fermented foods onto your plate.

ENABLER INGREDIENTS

Prebiotics		Fermented/ Cultured	Dairy Fermented/ Cultured	Non-Dairy Fermented
Artichoke	Jicama	Fish Sauce	Kefir	Sauerkraut
Asparagus	Leeks, Onions	Miso	Yogurt	Kombucha
Banana	(Emphasized	Soy Sauce		
Burdock Root	Here As	Tempeh		
Chicory Root	Prebiotics)			
Dandelion Root	Whole Wheat			
Garlic				

Chapter Thirteen: In the Kitchen with Enabler Ingredients

In the kitchen there are many ways to incorporate naturally fermented foods and prebiotics. I am a fan of creating your own ferments (loads of ideas and recipes to follow!) so you control the processes and you know how old, and therefore how alive, your products are. Whether it is fermented relish, pickles, beverages, or yogurt, making your own will always be superior to store versions because *you* control the ingredients, their freshness, and the age of your finished product. I like having this kind of power over what I eat. That being said, we are fortunate that fermented foods have come into their own these days. If you don't want to try your hand at making them, you can buy them. Many local producers have sprung up, and you can taste and find their products at farmers markets or at your local co-op or in the refrigerated natural foods or dairy section of your grocery store. To be sold retail, these products must be produced in licensed food production facilities and adhere to food safety regulations. Be sure the label says fermented or cultured. Some brands such as Bubbis (sauerkraut, pickles), Sunja's (kimchi), Miso Master (miso paste), Green Valley Organics (kefir), to name a few, have wide distribution and are more available across the country. Regardless of the product you buy, read the label to learn how the product is made and which ingredients are in the product.

On that note, if you are purchasing a fermented vegetable or fruit product, the only ingredients on the label should be some combination of *raw* ingredients, salt (for savory products), herbs, and spices. There should be absolutely no preservatives or sugar, which is why these products are found in the refrigerated sections of stores. Remember that fermented products extend beyond sauerkraut and pickles to include yogurt, miso, tempeh, and kombucha, as examples. There are a handful of traditionally fermented products—soy, tamari, teriyaki and fish sauces—which are found in the shelf-stable areas of the store. However, these products are produced in huge quantities, often using chemical hydrolysis to speed up the fermentation process, with coloring and flavoring sometimes added back in to reproduce the traditional flavor. Again, *read* the label to know what is in the product. Mass-produced fermented products are not the same as their naturally fermented counterparts. Use them as a condiment, or as directed in your recipes, but don't kid yourself that they are same as a living product you would make yourself.

Easy Ways to Get Enablers on Your Plate

For many of you, the idea of making your own probiotics is just too daunting. And what about those prebiotics? What's a person to do? So if this is you, here is a list of ten ways you can get store-bought *ferments* on your plate, followed by ten ways to ensure you get *prebiotics* on your plate. For the adventurous among you, speed read through this section to the recipes.

Ten Ways to Get Store-Bought Fermented Foods onto Your Plate

1. Add a spoonful or so of your latest favorite fermented product to your lunch or dinner plate every day. Soy sauce on your rice, fish sauce on your stir-fry veggies, you get the idea.

2. Kimchi, sauerkraut, or similar products taste fantastic with scrambled eggs. Kid you not!

3. Build a bowl and add a spoonful of your fermented relish as the garnish.

4. Eat plain yogurt. (Yes, you can ferment non-dairy milks, but they require a specific culture to achieve a probiotic result). Sweeten with a little honey (honey contains the bioactive, chrysin). Add your choice of fruit and nuts or seeds from the Genomic Kitchen Ingredient Toolbox.

5. Make a *raita* or *tzatziki* sauce, both yogurt-based sauces, to serve with salads or veggies. *Tzatziki* stems from Greece and traditionally uses cucumber, dill, and lemon with yogurt. *Raita* also uses a yogurt base and usually includes cucumber, but it can also be made with different vegetables. *Raita* may feature a spicy element such as cayenne and/or ginger.

6. Drink a glass of kefir or kombucha. Even a *small* glass counts!

7. Create a salad dressing using miso.

8. Make miso soup using the refrigerated store-bought, naturally fermented product as the base.

9. Substitute tempeh for meat. Cut a package into one-inch squares. Steam for ten minutes. Toss in a marinade of your choice (sesame-soy-ginger!) then stir-fry or bake. Loads of recipes out there. Just grab one and try it.

10. Go to *The Genomic Kitchen's* Enablers Board on Pinterest for tons of ideas and recipes!

Ten Ways to Up the Prebiotics in Your Diet (and Feed Your Probiotics)

1. Grill or roast asparagus. (Remember, grilling time will depend on thickness, so keep an eye on 'em.) Toss in olive oil before cooking. Serve with a sprinkle of salt, pepper, and whichever herbs are in season.

2. Use fresh or canned artichokes in your salad.

3. Make an artichoke dip. Use a white bean or hummus-based recipe. Throw in 1-2 store-bought artichokes from a jar. Easy.

4. Grate Jerusalem artichokes onto a salad, cook them like a potato (peeled/unpeeled) and mash along with other roots, or roast them along with your roots.

5. Eat a banana, or add it to a yogurt. (See probiotics list.)

6. Add banana and wheat germ to a smoothie or your yogurt.

7. Fresh leeks, onions, garlic: use them as foundations in your cooking.

8. Substitute chicory root coffee for real coffee on occasion. I really enjoy Dandy Blend which is a mix of chicory *and* dandelion roots (another prebiotic).

9. Grate jicama into salads. Jicama is a great addition to my Root Vegetable Slaw from Chapter Six.

10. Burdock root: add raw, grated to salads or thinly sliced, sprinkle with a little salt, drizzle olive oil, scatter a chopped herb or two.

In the Kitchen: Making Your Own Fermented Foods and Recipes

Let me begin this section with some advice and a recommendation: if you are a beginning food fermenter, start with a simple recipe such as my sauerkraut. Understand the recipe process, then, if you enjoy your creation, branch out. These days, there are many classes offered by local cooking schools and kitchen-supply stores where you can learn the basics and then get creative. If you are an online learner, there are several websites where you can watch a video, read myriad articles, and download tons of recipes. I have kept my recipes very basic in this book because there are so many resources out there. My go-to online sites for learning are (in no preferred order):

1. Cultures for Health has wonderful articles, recipes, and videos.

2. Body Ecology has articles, recipes and a chef-led Body Ecology Cultured Fermentation course. Look in Body Ecology for deeper learning and inspiring courses.

3. Nourished Kitchen features articles, recipes and how-tos, all organized by fermentation technique and beyond.

If you prefer to read and follow along in your own kitchen, the following books are my go-tos for recipes and ideas. More books are coming out every year, so keep your eyes peeled.

1. The Nourished Kitchen by Jennifer McGruther.

2. Fermented Foods for Health by Deirdre Rawlings, ND.

3. Fermented Vegetables: Creative Recipes for Fermenting 64 Vegetables & Herbs in Krauts, Kimchis, Brined Pickles, chutneys, Relishes & Pastes by Kirsten and Christopher Shockey.

Recommendations

Making your own fermented foods is not dangerous! Here are some simple recommendations to ensure that you produce a flavorful, health supportive product that is nourishing, safe to make, and delicious to eat.

1. Use the freshest ingredients possible. Buy locally and in season to ensure optimal freshness and nutrients.

2. Wash all your ingredients before preparing them.

3. Use sparkling clean, sanitized jars. To sanitize, fill your jars with boiling water and let them stand for a couple of minutes. Carefully pour water

out the jars, then place them in a 350°F. oven to dry. Carefully remove them and cool. You can then use them, or cap them with their lids to use when you are ready. Alternatively, run your jars through a normal dishwasher cycle, provided it includes the heated dry cycle. Then be sure to use your jars right away, once they are cool.

4. You don't need fancy fermentation equipment. If you want to use it, then buy it. If you don't, then traditional **Ball jars** (or sealable glass jars) are perfect. The company also makes simple, inexpensive equipment for handling jars and lids, wide-mouth funnels, and other canning gadgets.

5. Once the product is fermented and ready to eat, be sure to store it in a cool, dark place for the recommended time (according to the recipe). If your recipe says refrigerate the fermented product immediately on opening, do it. Once you open a fermented product, it must be kept in the refrigerator.

Recipes

Finally, you've reached the how-to section! What I have done here is feature recipes that showcase not only the nutritional power of fermented foods, but the different ways you can create them using several basic techniques. You can make your own fermented foods completely from scratch, or mix a store-bought lacto-fermented product with seasonal ingredients to put delicious food on the table that serves you, your gut, and your overall health. Let's get started.

Sauerkraut

Consider this a master recipe. Unlike the other recipes here that follow a more traditional recipe format, I am going to talk you through this one step by step. Let's keep it easy and start with the ubiquitous cabbage and a bit of coarse (large grain) sea salt. If you prefer a more traditional taste, you might want to have juniper berries or caraway seeds on hand, as well. In this recipe we are creating a salty brine that is the starter medium for fermentation.

The brine creates an environment in which anaerobic bacteria will thrive. In the first stage of fermentation in your jar, *leuconostoc* bacteria produce carbon dioxide and lactic acid. Both products are the result of the oxygen-free environment you have created by keeping your cabbage submerged in brine. When the environment becomes too acidic in the jar, it favors our gut-friendly beneficial species which start to multiple and thrive. *Lactobacillus* is the bacteria that gives sauerkraut its characteristic tanginess. This recipe uses the massage method to create the brine needed for fermentation.

Massage-Process Sauerkraut

PREP: ABOUT 30 MINUTES TOTAL.

MAKES ABOUT 1 TO 1½ QUARTS.

DO THIS

1. Grab one small head of cabbage, about three pounds. I like the Savoy cabbage, the one with the with deep dark green leaves that are curly and flopping out at the edges. Cut the cabbage in half.

2. Slice out the thick white core because it's a bit tough. Then put each cabbage-half flat side down on your cutting board.

3. Cut across the smallest diameter of the cabbage, creating strips of cabbage about ¼" across, like you are making strips for a slaw. Cut enough cabbage to create 8 tightly packed cups. Dump those 8 cups of cabbage strips into the largest bowl you can find.

4. Add 1 heaped tablespoon of the best coarse, uniodized (large crystal) sea salt to your bowl of cabbage. (Kosher salt is available at the grocery store.) Now start your workout.

5. For the next 5-10 minutes, grab the cabbage like you are wringing out water from a sponge and squeeze. What you are doing is massaging the salt into the cabbage to start the process of breaking it down. Keep on picking up handfuls of cabbage and wringing it. Keep going until you have a pool of salty water in your bowl, about ½ cup of liquid.

When you see this, you can stop wringing out your cabbage. If you like a more authentic sauerkraut flavor, add a teaspoon of juniper berries or caraway seeds to the cabbage mixture before you start wringing it.

6. The next step is where the magic happens. It is also the second part of your workout. Take a clean, 1-quart glass Mason jar (wide-mouth is easiest) and stuff about a cup of cabbage into the jar. Take a large wooden spoon and really start pressing the cabbage down. You want to keep pressing until the cabbage is *submerged* in the salty liquid. When you see this, it's time to add the next cup of cabbage.

7. Keep repeating this process until you have a jar full of cabbage and liquid. Stop adding cabbage when the liquid rises to within 1" of the top of the jar. Fermentation causes gas to escape, and you need to leave room for that expansion. If you have extra cabbage, you can repeat the process using another jar.

8. Now put the flat lid on the jar and *loosely* screw on the threaded collar. Keep the collar loose for now, and put the jar in a shallow bowl.

9. Put your jar on a counter or shelf in the kitchen, on a rimmed plate, away from bright light and heat. Leave it there for ten days. Yes, really!

10. During the ten days, you will see the cabbage change color and maybe some of the liquid escape from the jar. This is a natural process so don't be alarmed. Just make sure the cabbage is always covered by the liquid as this is important to the fermentation process.

11. After ten days, open the jar and taste the cabbage. It should have that characteristic pickled flavor, but still have crunch. If you like the taste, its time to tighten the lid and put your jar in a dark, cool storage place. Or, if you want to eat it right away, have a taste and then put it in the fridge.

Note: If you do use another jar and the mixture does not fill the jar, you can add a brine mixture to fill the jar to 1 inch below the top of the jar. Use Chef Catherine's brine formula from the next recipe.

Quick Lacto-Fermented Ginger-Lime Carrots

Chef Catherine Shamburger Brown is a beautiful, certified plant-based chef based in New Hampshire who is also pursuing a degree in nutrition. Her food is absolutely gorgeous and her recipes simple and divine. See for yourself at her site and follow her blog, "A Seat at My Table." In this delicious recipe, Catherine uses the brine method (versus the massage method) to create the fermentation environment. Takes less than 30 minutes start to finish, and is ready to eat in 3-5 days based on your personal taste preference.

PREP: ABOUT ONE HOUR.

MAKES 1 QUART IN A 1-QUART MASON/BALL JAR.

GRAB THESE

3-4 cups filtered water, purchased or from refrigerator dispenser

1 pound carrots, washed and dried

2 tablespoons fresh ginger, grated (about a 2" knob)

2 large limes, washed

1 tablespoon unrefined, non-iodized sea salt

DO THIS

1. Bring 1 cup of filtered water to a boil and dissolve the salt in it. Cool and add 3 more cups of cold, filtered water to the salt water. Stir with a clean spoon to combine. Use filtered versus tap water to eliminate and potential contaminants.

2. Meanwhile, spiralize or grate carrots to yield 4 packed cups. Put them in a **non-reactive bowl.**

3. Peel and grate the ginger and add it, along with its fiber, to the carrots.

4. Zest the limes, using a microplane or zesting tool and add to the carrots.

5. Squeeze the lime juice over the carrot mixture, extracting as much juice

as possible. Use a *clean* hand, tongs, or a spoon to combine.

6. Add the carrot mixture to your jar a large spoonful at a time. Pack each spoonful down tightly so that you have a compact product. Leave 1-2" headspace between the compacted carrot mixture and the very top (lip) of the jar.

7. Completely cover the carrot mixture with the cooled brine, leaving 1-2" of head space. Be sure carrots are completely submerged. Save any leftover brine to top up the jar if leakage occurs during the fermentation process, exposing the carrots to air. Remember, fermentation is anaerobic, so we keep all product submerged in brine.

8. Place the lid and collar on the Mason jar. Secure the collar only to the point that it can be easily turned or removed with one hand. Do *not* tighten it down. As the product ferments, gas is produced causing the liquid to expand and sometimes bubble out of the jar, a sign the process is working! Leaving the lid barely tightened allows this expansion to occur.

9. Label and date your jar, then place it in on a plate or tray and out of direct light and heat, but at room temperature. Your pantry is a good place. The tray will catch any of the liquid which might escape from the jar.

10. Leave at room temperature for three days, then give the product a taste. If you prefer a more pronounced flavor or tanginess allow the carrots to continue fermenting at room temperature for another 3-5 days, tasting intermittently. When you are satisfied with the flavor and texture, tighten the lid and refrigerate or store in a cold (not freezing) basement or root cellar. The carrots will keep in the refrigerator for at least nine months—if you don't eat them up first!

Note: the warmer the room temperature, the quicker the fermentation will happen. Ideal ambient temperature is 60-70° F. If you are concerned at all about your room temperature, just leave the jar out for 24-48 hours to begin the fermentation process and then store the jar in the refrigerator, cold root cellar, or basement. It will take longer for the veg to acquire a nice tanginess, but it will eventually get there.

Asian-style Pickled Vegetables

I met my friend and chef Kai Peyrefitte while working on an innovative culinary project in Southern California. Kai is a passionate chef, with deep expertise in Asian-style cuisine. While working in the test kitchen for our project, Kai regaled us with beautiful bowls of pho, from-scratch miso soup, as well as a chocolate ganache made from probiotic-rich yogurt. Kai is an avid believer in the importance of gut health and creates recipes that treat your palate as well as your gut! I could not write a book without including at least one of his inspirations.

PREP: 25 MINUTES.

YIELD: 1 POUND.

GRAB THESE

1 ounce organic Soy Sauce

1 ounce honey (Manuka if possible)

1 ounce ginger, minced

1 ounce garlic, minced

1 tablespoon chili flakes

1 pint water

2 ounces unfiltered vinegar (such as Bragg's)

1 ounce salt

4 ounces* carrot, thin sliced

4 ounces* red onion, thin sliced

4 ounces* red bell pepper, thin sliced

4 ounces* celery, thin sliced

   4 ounces is approximately a heaped half cup

DO THIS

1. Add the first 8 ingredients to a non-reactive bowl.

2. Mix until ingredients are well combined and the salt has completely dissolved.

3. Gently fold in the vegetables.

4. Completely cover with a lid and refrigerate for 5 days.

5. Open the lid every couple days to release gas that may be forming.

6. After 5 days, place the pickled vegetables in a sealable container and enjoy.

Tip: this is a cold (versus room-temperature) fermentation. If you see bubbles forming in the liquid, that is good! That means good bacteria is growing.

Fermented Red Cabbage and Beet Salsa

I teach a lot of online courses for clinicians and chefs. In one course, I noticed Donna Gates had registered. *The* Donna Gates, I wondered? Donna Gates is a lifelong student, author, teacher, and recipe innovator and, most importantly to me, also a guru in gut health. Donna is the woman behind Body Ecology, the website I cited earlier as a great source of information, recipes, and tutorials about fermentation. I thought nothing more about it and went on to teach taught the course. Then I noticed Donna in another course. I was surprised. What could *I* teach Donna Gates I wondered? She truly pioneered the movement that connects the role of fermented foods to gut health and healing. Having her in my courses was both daunting, but also refreshing as she openly shares her vast knowledge with fellow students. I asked Donna to share one of her recipes and was delighted when she agreed.

For this recipe, Donna created a sweet brine for a very specific reason. Always insightful, she recognizes that beets are extremely high in oxalates which can be problematic for some people. And yet, beets are nutritionally valuable, as you learned in Chapters

Seven and Eight, on the Influencers. By fermenting beets using her sweet brine formulation, Donna was able to remove *all* the oxalates in the beets and still give you full access to their nutrition potential. (A University of Nebraska lab validated this finding.) Oh, and if you look at the other ingredients in this salsa, you'll find them in the GK Ingredient Toolbox. Thanks, Donna! We now have a fermented product that speaks to our genes, our biochemistry, and our gut. (Note that this recipe does include proprietary products from Body Ecology which you would want to buy to make it the exact way Donna makes it. Body Ecology products are available through Amazon or Body Ecology's website if you are interested in purchasing these exact ingredients.)

PREP: 40 MINUTES (ADD EXTRA TIME IF YOU ARE HAND GRATING INGREDIENTS). MAKES APPROXIMATELY 3 QUARTS IN 3 1-QUART MASON/BALL JARS.

GRAB THESE

FOR THE BRINE

3 red apples

2 teaspoons Celtic or other coarse sea salt

1 teaspoon Body Ecology EcoBloom powder

1 packet of Body Ecology Culture Starter

6 capsules Body Ecology Ancient Earth Minerals

4 cups water

FOR THE SALSA

3 heads of red cabbage, outer leaves and white core removed

3 large beets, peeled

1 large bulb of fennel, outer leaves removed if soiled

1 bunch kale

1 large bunch of cilantro or more to taste

3 red peppers

DO THIS

1. Place all the sweet, enriched brine ingredients in a blender or small food processor and process until well blended. Set aside.

2. Using a large food processor, mandolin, or grater, shred the cabbage, beets, and fennel and place in a large bowl.

3. Remove the stems from the kale (use a knife or tear the stem out with your hands), then roughly chop it into 1" pieces.

4. Rough chop your cilantro, discarding the long stems.

5. Cut the red peppers in half. Remove the core and seeds and dice into ½" inch pieces.

6. Using your hands, roughly toss the kale, cilantro, and red peppers into the cabbage mix.

7. Pour the starter culture mixture over the cabbage mix and mix well with a large spoon.

8. Tightly pack your sterile Mason jars with the cultured vegetable mixture, leaving about 1½" at top.

9. Seal each jar, but not too tightly (see recipes above). Run under hot water and wipe clean.

10. Ferment for 5-10 days at room temperature, in a location not exposed to direct heat or light.

Note: The longer you ferment, the more distinctive the flavors and the better the medicine for you and your gut. The distinctive flavor of lacto-fermented foods comes from the strains of beneficial bacteria produced through fermentation.

Simple Salad with a Fermented Kick

I am a great believer in sharing great recipes and great work. This super simple delicious salad was shared by my great colleague, Gay Riley. Gay is a functional-medicine nutritionist in Dallas, specializing in nutrigenomics. Her delicious, easy-to-make salad features fresh ingredients that she creatively mixes with a store-bought fermented pickle product—and uses its juice, too! Downsize this salad to serve individual portions, or triple it to feed a crowd!

PREP: 30 MINUTES.

SERVES 4-6.

GRAB THESE

2 cups peeled and diced Persian cucumbers

2 cups diced cherry tomatoes

1 cup diced red onion (or more if desired)

1 cup diced lacto-fermented dill garlic pickles (a locally produced product if you can)

4 tablespoons pickle juice (from the pickle jar!)

Few drops honey

1-2 teaspoons chopped mint

DO THIS

1. Place all the ingredients except honey and mint in a large serving bowl.

2. Gently fold together, adding honey and mint to taste before chilling for one hour. Enjoy!

Optional additional ingredients: because this is such a simple salad, it lends itself to on-the-fly additions. Add a ¼ cup (or more) of pitted, sliced black olives (whichever variety you prefer) or diced avocado, crumbled feta cheese—or all of them!

Note: The leftover juice in any naturally fermented product is one of the best tonics you can drink. Don't throw it away! Either drink it, or add it to your next batch of fermented product. It serves as a seeding medium, adding probiotics into your next batch of product.

Yogurt Sauces

Yogurt is one of the easiest fermented products that you can use to seed your gut. If you like yogurt, then these simple cucumber-based recipes with different twists can add a refreshing taste to hearty salads, seafood, chicken, or even serve as a dip with some fresh veggies. I recently made a *raita* and paired it with a salad made of grilled cabbage and fresh cucumbers after getting the idea from a culinary magazine. So pair either of these recipes with grilled or roasted vegetables, too. I recommend you buy organic yogurt, preferably a Greek- or thicker-style yogurt versus a fluid one.

Tzatziki

PREP: 15 MINUTES.

MAKES 1½-2 CUPS DEPENDING ON SIZE OF CUCUMBER.

GRAB THESE

1 cup whole milk Greek yogurt

1 *English* cucumber, peeled, seeded, cut into ½" dice

2 medium cloves garlic, smashed and finely minced

2 tablespoons extra virgin olive oil

1 teaspoon lemon zest

1-2 tablespoons fresh lemon juice

¼ cup chopped fresh dill

¼ teaspoon freshly ground sea salt

Cracked black pepper to taste

Pinch cayenne or smoked paprika

1. Place the yogurt, cucumber, garlic, olive oil, lemon zest, and juice in a bowl and gently combine.

2. Fold in the dill, salt, and a few grinds of black pepper. Taste and adjust seasonings, adding more dill or lemon juice if preferred.

3. Place in a serving dish and chill at least 30 minutes to allow flavors to meld.

4. Serve with a pinch or so of cayenne or smoked paprika.

Note: You can change this up in many ways. Try adding 2 tablespoons of chopped mint or basil to the recipe, or 1 tablespoon of chopped fresh thyme. Or just add all three of these herbs, which are rich in bioactives. If you want to enrich the recipe and create a thicker sauce, put everything in a small food processor and add 4 ounces of soft goat or feta cheese. Divine!

Raita

Similar to *tzatziki*, we use a yogurt-and-cucumber base to make *raita*, adding flavor nuances through the use of herbs and spices. I have seen many different variations including ginger, cumin, coriander, cayenne, hot chili peppers (fresh), cilantro, and mint. Here's my version of *raita*. Just play around and make it your own.

PREP: 25 MINUTES.

MAKES 1½+ CUPS DEPENDING ON SIZE OF CUCUMBER.

GRAB THESE

1 cup Greek yogurt (I prefer full fat)

2 small cucumbers, peeled and rough chopped (I mean the small ones, or use ½ an English cucumber)

¼ cup cilantro, stems removed and minced

1 medium jalapeño (or a small serrano chili), halved, seeded, and rough chopped

Juice of ½ lime

1 large clove garlic, peeled and smashed

1" piece fresh ginger, peeled, smashed, and roughly chopped

½ teaspoon coriander

½ teaspoon cumin

Two pinches cayenne (optional)

¼ teaspoon kosher salt (more to taste)

DO THIS

1. Place all ingredients in a small food processor and blend until combined.

2. Taste and adjust salt, cayenne if needed.

3. Chill for at least 30 minutes to allow flavors to meld.

Note: if you strain the yogurt for 30 minutes using a fine mesh strainer over a bowl, you will create a thickened yogurt "cheese" product. Using this as a base for the raita creates a thicker sauce. You can always add back in a little more yogurt if it is too thick.

Something Sweet

No book with recipes is complete without something a little sweet! As you know, this book is not a cookbook, rather a book about how to choose and prepare ingredients that give a helpful boost to your genes. In this chapter I have focused on recipes that feature the fermentation to produce probiotic rich bacteria. But beyond bacteria, we also need to support the gut with foods rich in fiber to help your eliminate waste from your body. Like so many of the chefs I work with these days, **Chef Pam Florence** is deeply interested in the science of culinary genomics, and how she can apply it in her work with clients. When I asked her if she had fermented anything recently, she said, no not right then, but she did have a lovely pickled plum recipe. "Send it to me," I said! So here is Pam's beautiful pickled plum recipe. It is not a fermented recipe, rather a recipe rich in fiber and deliciousness which you and your gut will enjoy.

Pickled Plums

PREP: 15 MINUTES. COOK: 10 MINUTES.

YIELD: 1½ QUARTS.

GRAB THESE

½ pound, about 3-4 medium plums, firm but ripe black or red plums, rinsed, pitted and quartered

¼ cup red wine vinegar

¼ cup good quality balsamic vinegar

½ cup water

⅓ scant cup raw honey (or more if you like a sweeter pickling liquid)

1-2 whole star anise

3 whole cardamom pods

1 cinnamon stick

6 pieces of fresh ginger, peeled and sliced into rounds

⅛ teaspoon sea salt

DO THIS

1. Make a few slits on the skin of each plum with a sharp knife. Place the plums in a clean, glass (heat safe) jar and set aside.

2. Place the vinegars, water, honey, star anise, cardamom, cinnamon stick, ginger, and sea salt in a small non-stick or non-reactive saucepan. Simmer the ingredients over a low heat for about 10 minutes until the liquid is reduced to about ½ cup.

3. Pour the hot pickling liquid over the plums in the glass jar, leaving ½" of room at the top.

4. Let the plums cool, cover, and place in the refrigerator for at least 2 days, or up to 7 days.

And One More Thought for the Road

With over more than 20 years in nutrition practice and education, I have learned that we always have to start with people's own food experience. As I say, you can't expect people to eat kale when sweet corn has been their "go-to" as a vegetable (ok it's a grain). The point here is that fermented foods are very new to many people, and they present a scary unknown. Given the love affair so many Americans (and others) have with ketchup, why not start by replacing the store bought one with your own? As always, there are lots of recipes out there on the web, so you can just search for "fermented ketchup," or go to our Pinterest boards and you'll find a selection of ketchup recipes on the Enabler Board.

Chapter Fourteen: Cooking and Eating with the Genomic Kitchen Ingredient Toolbox

A Deeper Dive

At the beginning of this book, I took you on a panoramic and gastronomic journey to the Italian Dolomites in Northern Italy to share a meal with you, if only vicariously, eaten at AGA Restaurant, a one-star Michelin establishment. That meal was the first step on a journey to opening your mind to the ideas of culinary genomics and the Genomic Kitchen. I will begin this final chapter with yet another food journey, this time to a much less glamorous village and meal, but one just as important to bookend our journey.

In the fall of 2017, I was again visiting Northern Italy. Many years ago, I traveled down to the French Riviera from Vicenza, down to Genoa, and then east on the highway to France. It takes about five hours. At about the four-hour point, hunger struck, no doubt provoked by the luscious surroundings of the Italian coastline where indeed, mountains flow down to meet the ocean. Traveling with a friend, we came across the exit for Finale Liguria and, instead of heading to the town, we headed north into the mountains. About five minutes later we parked in the shuttered and silent hamlet of Gorra. Now, Gorra was not closed for business. It was closed for lunch, except of course, if you own the local *trattoria*. In that case "closed for lunch" means it is time to earn your living.

Nosing around for the faintest sign that at least one person was serving lunch in this tiny place, we happened upon an *osteria*, which in Italian generally signifies a simple, inexpensive experience. So we wandered into Ristorante Osteria dell'Agorà, hoping that someone was in the kitchen. Sure enough, the chef was alerted that someone (us) wanted something to eat. Miraculously, three people were available to guide us to an empty table in an equally empty restaurant. We by-passed the ancient olive mill residing in the entry way and descended three steps into a whitewashed stone room with windows opened to the mountains. This was indeed our lunch place. Wine lists are never required in these places, because inevitably the wine is local, it's Italian, and it is therefore good. After the wine appearance came the menu, which also did not need navigating because the daily specials are what one chooses. Why? Because in this part of the world, they usually reflect a combination of the day's catch—we could see the ocean—in combination with whatever is in season and freshly acquired that day.

A risotto featuring fresh squid and other small fish appeared rapidly, along with one of my favorite dishes, a simple combination of potatoes and green beans tossed with a pesto and fettuccine, known as *Trofie pasta Liguria*. Note the pesto is aptly named Pesto Genovese.

The reason I tell you about this place is because the cuisine was startling simple, reflecting the uncomplicated way of life of this part of the world. Let's be clear, these people work hard, but the traditions of the table and the nourishment that is delivered there are a central part of life. In common with AGA, that award-winning restaurant in the Dolomites, the chef uses seasonal ingredients. At AGA, chef Oliver leveraged his extraordinary culinary talent to create gastronomic art on a plate, capturing the essence of the season with local high mountain ingredients. At Ristorante Osteria dell'Agorà, the dishes were equally as flavorful, yet refreshingly simple. This tiny restaurant menu has not changed in hundreds of years. The dishes simply work. They are not complicated, reinforcing the theme I want to have resonate for you; namely, that cooking does not need to be complicated to be flavorful and nourishing. Green beans and potatoes with a simple pesto of basil and olive oil, garlic, and nuts is not hard to make and not expensive, when made in season.

As I read this paragraph to a colleague, he noted two more things that are important to share. First, that at both AGA and the *osteria* in Gorra, what arrives on your plate is both nourishing and flavorful, regardless of price. You can pay a lot (AGA) and you can pay a lot less. Generally, you still get a delightful meal. The second point is that in both of these restaurants, the turning of the table is not important. By this I mean, the table you sit at is yours until you vacate it.There is no expectation that you finish your meal on a timetable and dutifully depart, so you can enjoy your food versus shoveling it down. No one is hovering to give you a bill. Instead, your table becomes your kingdom, a place to meet, raise a glass of water or wine, acknowledge the people you are with, and the food that you will eat. Food is information, but it also facilitates communication.

Now in case you think I spend my life searching for unique restaurants in out-of-the-way places in Europe and beyond, I don't. Most of the time I live and work in the United States. My work in the US has taken me to fine dining restaurants, but more importantly, to hundreds of small towns and communities across this great country. Over the past 20 years, I have watched tiny farmers markets flourish, from La Grande, Eastern Oregon, to Bismarck, North Dakota, to Frederick, Maryland. I have watched as local, state, and national initiatives have made locally grown food available to both young and old in need and to those with financially few needs. I have seen fermented foods move from "something your grandma did" to becoming a "must-taste" stall at markets. Artisanal bread and local cheeses now captivate mouths and minds, while locally raised meat takes its rightful place in local communities. You now find menus touting locally sourced seasonal ingredients with pride at tiny local diners and upscale restaurants alike. And backyard gardens are a source of many conversations and lots of local call-in radio shows.

In the Driftless region of Wisconsin next to the Mississippi River, where I currently live, the glaciers bypassed this area and left the soil as black as coal and mineral rich. This region of blessed topsoil between Madison (to the south) and Minneapolis (to the north) is home to America's corn and soy fields. Next to those fields are grasslands as green as the Emerald Isle itself. And with that grass has come a new era of cheese-making rivaling the selection I grew up with living in France. In the local farmers markets, growers are introducing curious shoppers to the taste of an authentic heirloom tomato, huge bunches of herbs ripe for a pesto or a flourish of flavor on tonight's meal. The Driftless region is relatively poor economically, but rich in soil and the incredible nutrients that the soil delivers to food. In turn, these nutrients—when eaten— give our bodies the fuel to stay healthy. Soil is instrumental if food is to have nutritional value. Consuming good food—particularly those foods in our GK Ingredient Toolbox—well-prepared and with reasonable consistency translates to optimal functioning and performance of our genes. The net result: our genes produce the proteins that communicate with the biochemical pathways that define our health. Food does not need to be complicated. It does need to be good. You need to buy good food and prepare it simply.

Cooking with Food that "Talks" to Your Genes

Many of you have read this entire book thus far and picked up lots of ideas and tips along the way. That being said, I also know a bunch of you have speed-read or page-flipped and landed at this point. You wanted to get to the "quick download," the "411" of how to do the food-gene-cooking-relationship thing. Well, if you are the skimmer, then I have a few tips and more new ideas for you here. Sure, you can run with the ideas in this final chapter and make some good food progress, but you are going to miss the mortar that holds the bricks together. So, sure, read this chapter, but then work backwards. It's worth your time.

The Complete Genomic Kitchen Ingredient Toolbox

Be honest! Did you race to this last chapter so see what the comprehensive ingredient looks like? If so, here it is just below. I have organized the Genomic Kitchen Ingredient Toolbox by food group. To see which ingredients are in each food group, use the color of that group up top to navigate you to the corresponding ingredients below. As you know, I carefully unpacked the Genomic Kitchen Ingredient Toolbox over the course of previous chapters. In each chapter I have given you individual summaries of which ingredients apply to the respective M, I, S, or E groupings, and why. Here you see everything all at once. If this is too busy for you, just go back into the chapters and refer to each individual chart which you can also download from my website. You can also use the same link to find and download the Genomic Kitchen Ingredient Toolbox, or the ingredients organized by MISE.

Genomic Kitchen Ingredient Toolbox

FRUIT

Citrus
PINK GRAPEFRUIT

Fruits That Are Like Vegetables
AVOCADOS, TOMATOES (COOKED)

Berries
BLUEBERRIES, BLACKBERRIES, ELDERBERRIES, RASPBERRIES, STRAWBERRIES

Other
APPLES, GRAPES (RED), MANGO, ORANGES, POMEGRANATE, WATERMELON

VEGETABLES

Cruciferous
LEAFY
ARUGULA, BOK CHOY, COLLARDS, DANDELION GREENS, KALE, MIZUNA, RADISH LEAVES, TATSOI, TURNIP GREENS, WATERCRESS

HEARTY
BROCCOLI, BRUSSELS SPROUTS, CABBAGE, CAULIFLOWER, KOHLRABI, RADISHES, RUTABAGA, TURNIPS, WASABI

Alliums
GARLIC, LEEKS, ONIONS, SHALLOTS

Other
BEETS, CARROTS, CORN, FENNEL, PEAS, PEPPERS, SPINACH

LEGUMES
BROAD BEANS, CHICKPEAS, LENTILS, PEANUTS, SOYBEANS

NUTS & SEEDS
CHIA, BRAZIL NUTS, FLAXSEED, HEMP SEEDS, SESAME SEEDS, SUNFLOWER, SEEDS, WALNUTS

ANIMAL PROTEIN
CHEESE, BEEF, LAMB, PORK, POULTRY, WHOLE EGGS, WILD GAME

SEAFOOD
ALBACORE TUNA, ANCHOVY, HERRING, MACKEREL, MUSSELS, OYSTERS, PACIFIC HALIBUT, SALMON, SARDINES

HERBS
BASIL, CHIVES, CILANTRO, DILL, LAVENDER, LOVAGE, OREGANO, PARSLEY, ROSEMARY, SAGE, THYME

SPICES
GINGER, HOT CHILI, PEPPERS, TURMERIC

OTHER
BREWERS & NUTRITIONAL YEAST, GREEN TEA, HONEY, MUSHROOMS, OLIVE OIL, RED WINE, RYE, SEA VEGETABLES, QUINOA, WHEAT GERM

GUT SUPPORT

Probiotic Rich Fermented/Cultured
FISH SAUCE, MISO, SOY SAUCE, TEMPEH

DAIRY FERMENTED/CULTURED
KEFIR, YOGURT

NON-DAIRY FERMENTED
SAUERKRAUT, KOMBUCHA

Prebiotic & Fiber-Rich
ARTICHOKE, ASPARAGUS, BANANA, BURDOCK ROOT, CHICORY ROOT, DANDELION ROOT, GARLIC, JICAMA, LEEKS, ONIONS, WHOLE WHEAT

The Genomic Kitchen Ingredient Toolbox

If you have read the book thus far, you already have lots of tips, strategies, and sample recipes to help you incorporate these ingredients into simple, delicious meals. But for some of you, I know that is simply not enough. Here you stand armed with science and a bunch of ingredients, and you are still not sure what to do. You are not alone. Therefore, this chapter is dedicated to helping you organize your thinking about ingredients and recipes to bring the Genomic Kitchen into your home.

Here's how we're going to do this. We'll start with those vegetables and fruits which you *know* are the gateway to key aspects of that all-important food-gene relationship. After that, I'll introduce you to a Culinary Framework to help organize your thinking around which recipes to choose and how to put these ingredients on your plate in easy ways. I'll finish with a Quality Assurance Tool (yes, really!), that helps you cross-check your recipes at home and dishes you choose when eating out. So, several useful tools to share with you. Let's dive in.

The Road to Broccoli is Hard with a High Fructose Corn Syrup Palate

Let me share this advice, product of my roots as a kid in Europe, and then cultivated as a professional working in nutrition science for many years. Here's one thing we know about most individuals' relationships to food, their tastes, and preferences. *The food you were exposed to in your early life shapes your food choices today.* To put this in perspective, if you were one of those lucky kids whose parents grew veggies and fruit, or they cooked from scratch regularly in your kitchen, or you got the chance to eat out a decent restaurants, the chances are you have had the opportunity to cultivate a wide palate. A wide (diverse) palate means one that is open to lots of different food, textures, and flavors. And a wide palate means you are therefore exposed to a plethora of nutrients, and subsequently lots of food information. This does not mean all that information is or was great, but with a wide palate, you probably got a good dose of vegetables and fruit in the mix. If your experience was not this colorful, maybe your palate is not so wide. Also, if you were a kid or grew into adulthood with illnesses that prevented you from multiple food options, that too shaped your palate. Sometimes food hurts when it creates reactions and auto-immune responses, like those of us who suffer from allergies and sensitivities to seafood, peanuts, even all gluten.

So why do I share this? Because getting the right food information starts with what goes in your mouth. And what goes in your mouth should be the best information you can provide for your genes. Sometimes this is easier said than done. A phrase I often use is: the road to broccoli is hard when you have a high fructose corn syrup palate. What this means in daily life is that I can give you all the great scientific reasons to eat broccoli, rutabagas, fermented cabbage, and blueberries, but if your palate has only been exposed to the same ten foods, which don't include broccoli or blueberries, then it is possible that no amount of science will convince you to suddenly start eating those broccoli florets and blueberries. As you now know, integral to how well your body functions are vegetables, fruit, herbs, and spices. Bad news for you plant shirkers. Now, if you are one of those people who slinks past the produce department, or someone in your family does, I have a strategy for you. It is based on what fruit and veggies *taste* like versus what their nutritional (scientific) value is. This way you can buy produce based on flavor familiarity and expand your palate and nutrient information from this unique perspective. Having tested this concept with thousands of people, I can vouch that it will work for you, your partner, kids, and grandkids too.

Acquiring a Taste for Food You've Been Avoiding

Many years ago, I was presenting a tasting exercise at a conference for food and culinary professionals. It was a simple active exercise that involved tasting ingredients that were all cut to the same dimension and looked alike. The goal of the exercise was to introduce people to a new way of approaching food, namely choosing vegetables and fruit by what they taste like, and not based on their nutrient content. The premise of this exercise: with great food comes great nutrition. I learned this idea from a colleague long ago who said this, "You have to put food in people's mouths to open their ears. You can't download flavor and you can't eat handouts."

When It Comes to Learning—and Liking—a New Taste, It's Tricky!

That colleague, Mark Mulcahy, is a produce expert whose work involves training staff in grocery store produce departments around the US. The training has two goals. First, to create attractive produce displays that invite people to buy. Second, to train produce staff in the importance of their roles. As Mark told me, he advises produce staff that how they describe produce to people is critical. And you have to know what you're suggesting in terms of ripeness, preparation, and the actual taste of the item. He said if you ever sell a piece of produce to someone that does not deliver on the flavor and texture that you promise, they will never buy that produce again. Opportunity lost, maybe for a lifetime. I tell my fellow clinical colleagues the same thing. We rarely get a second chance to make a first impression with our advice. Get it right and make it enticing!

Back at my produce-tasting presentation. At the end, a lovely lady came up to me and said "I am Deborah Madison, and I want to do what you're doing." Deborah is a James Beard award-winning cookbook writer and founding chef of the famous Greens Restaurant in San Francisco. Once I picked my jaw up off the floor,

I knew exactly what project I wanted Deborah to work on, and that project is the work I want to introduce you to next.

You see, for many years, the nutrition and health community has presented food in terms of their nutrient-to-health value. Calcium is good for your bones. Vitamin A is good for your eyes etc. OK, I am being simplistic, but read a health article and sure enough, you'll see that we promote food primarily in terms of its health value. Well guess what, if you're not sure what that gnarly looking vegetable or funky-looking fruit tastes like, who cares what it does for your health? Flavor and texture are what help people lean into food, not the nutrient-to-health equation (for the most part). It's almost like health is a secondary benefit once you get on board with taste. Another way to look at this is if you can encourage people to try a wide variety of fruit and vegetables, then the nutrient benefits come with them. So think flavor and texture first.

This being the case, I asked Deborah to use her brilliant culinary and botanical mind to re-imagine the produce department, one that is organized by flavor and not by nutrients. This required Deborah to set aside the fact that some produce items can sit out at store temperature and some need cooler refrigerator–like environments. So imagine that chilled lettuce could sit next to oranges just for the sake of this exercise.

Now for a brilliant chef, you would think this is easy. It is not. Acknowledging the many factors that impact the flavor and texture of a plant was a hard task. High temperatures or frost, lots of water or very little water, and the structure and nutrient composition of the soil can all shape the taste and texture of the food in your mouth. Alas, we had to put all of these variables aside and imagine an ideal taste profile for each item in the produce section of the store. The point of this work was to create a tool that helps you navigate the produce section and grow your palate one bite at a time.

One Bite at a Time

The idea of organizing vegetables and fruit through flavor (or what you perceive they might taste like) came from my stepfather, a UK wine importer who had specialized many years ago in single vineyard wine varieties and champagne from France. He was exporting these treasures to the UK. Let me clear. *We did not own a vineyard and we weren't making wine!* He once told me about the anatomy of wine. This entailed understanding that wine has a mouth, a head, and even legs to describe its structure. Listening to him, it immediately struck me that the wine business has grown around the world, not by nutrient labels or caloric value or health claims, but by this idea of pairings. If you are a chardonnay lover, I can present three tastings of chardonnay from different world regions, or even different vineyards in one state or appellation. They may taste very different from each other, but you will not be afraid to taste them because you know you like chardonnay. If you don't like wine, you can think of tasting three different oranges, or three different apple ciders. What I am driving at here is that the confidence to try new foods starts with knowing how that new food fits into your comfort zone of preferred flavor.

Let me give you an example, and then you can jump to the tool. Let's say you love sweet corn, but are an avid avoider of all things green, except green beans. From a flavor perspective, sweet potatoes or snap peas might be a logical "next step" vegetable to try because they fit into a sweeter flavor profile. Jicama also fits into sweeter, and for some people, beets, too, particularly roasted. Yes, I know that some of you don't like beets!! My experience has been that when we navigate people to new food choices based on flavor perception (tastes you prefer) versus nutrition science (tastes you *should* prefer), we can gradually open up your palate to new things. Remember, we eat *food*, not science information on handouts or in apps.

What's the Difference between Flavor and Taste?

Many hands have been wrung over this question, including my own. Well, I like the way the **Center for Smell and Taste** at the University of Florida explains it. The words "flavor" and "taste" are used interchangeably, just like the words lawyer and attorney, or solicitor if you live in the UK. Whereas attorney and lawyer do actually refer to the same profession, flavor and taste will never share the same name or meaning as our legal companions. Taste is something that occurs in the mouth and is a reflection of information passed from the food to the brain which is interpreted as sweet, *umami,* or bitter for example. While taste is mainly a mouth to brain connection, flavor brings in another dimension, namely aroma. Aroma is detected through the nose, and according to the University of Florida, accounts for up to 80 percent of how we perceive flavor, versus taste which contributes a mere 20 percent. So how something smells informs taste more than what happens when food touches your tongue. Who knew? And oh, if you have kids and they refuse to eat mushrooms or bananas, it may be a lot less about perception of flavor, and a lot more about texture, which I talk about here very shortly.

How Cooking Technique Enhances Food You Thought You Did Not Like

OK, I have always loved the earthiness of Brussels sprouts, cabbage, cauliflower, and the like. Growing up in the United Kingdom (UK), my dad grew a lot of vegetables which arrived in a mere matter of footsteps from the back garden to the table. Yes, they may have been steamed and served quite simply, but often on Sundays as is typical in the UK, they were served richly enveloped in a caramelized crust because they cuddled up to whatever meat was being roasted and served that day. This is good old-school British cooking at its best!

What made these earthy root and cruciferous vegetables so tasty was the art of roasting. Roasting allows the natural sugars (carbohydrates) in vegetables and fruit to caramelize, creating the sweeter, crusty texture that some of us love. Think about how in vogue kale and Brussels sprouts have become. Toss them in olive oil (yes I said olive oil!) and spread them out on a baking sheet. Roast for 12-15 minutes and have a taste. You have a nice, crispy, sweeter rendition of your vegetable and super tasty too. Roasting causes caramelization, or the Maillard Reaction (depending on what you are roasting). Both of these cause the chemical changes that result in that crusty flavor you have grown to love. Yes, indeed, you can have those veggies and love them too. Just change up the cooking technique once in a while and you will create veggie lovers out of the most timid of people. Roasted garlic anyone? Read more how the roasting technique changes texture and flavor here.

Flavor Families

On the next page you'll find my palate navigation tool for vegetables, otherwise known as the tool that will help you navigate to new vegetable choices that might fit both your palate and your flavor style.

How to Use This Tool

You can see that I have organized common vegetables into flavor categories. (Okay, some are fruits or grains that we serve as vegetables.) If you choose one category and pick out a few familiar veggies, you will quickly realize that they are united by a fairly similar taste. I already talked about sweeter veggies a little earlier, so let's take a look at the spicy category. If you have ever tasted fresh arugula or mustard greens, radishes, or watercress, you *know* they have that kick of heat. I call the spicy category the "tailgate" category, to describe all the people who swear they won't eat vegetables, but love to dip their snacks into hot salsa or dump sriracha all over their food at a tailgate or backyard party.

Vegetable Flavor Families

SWEET	BEETS, JICAMA, PARSNIPS, PEAS, THYME, TOMATO, SNAP PEAS, SNOW PEAS, SWEET POTATOES, WINTER SQUASH,
MILD/ NEUTRAL	BOK CHOY, BOSTON BIBB, CHARD, DAIKON RADISH, EGGPLANT, JICAMA, MIZUNA, NAPA CABBAGE, POTATOES, SPINACH, TATSOI, ZUCCHINI
EARTHY/ PUNGENT	BEETS, BROCCOLI, BROCCOLI RABE, BRUSSELS SPROUTS, CABBAGE (SAVOY), CAULIFLOWER, CELERY ROOT, COLLARDS, KALE, KOHLRABI, MUSHROOMS, PARSNIPS, RUTABAGA
GRASSY	ASPARAGUS, CELERY, CHARD, CUCUMBER, FENNEL, FRISEE, GREEN BEANS, MIZUNA, PARSLEY, ROMAINE, SHISO, SNOW PEAS, SPINACH, TATSOI
LICORICE	BASIL, ENDIVE, FENNEL
SPICY	ARUGULA, BASIL, CHILI PEPPER, HORSERADISH, LEEKS, MUSTARD GREENS, ONIONS, RADISH, TURNIP GREENS, TURNIPS, WATERCRESS
TART	LEMONGRASS, TOMATILLO, SORREL
BITTER	BELGIAN ENDIVE, CHICORY, CURLY ENDIVE, DANDELION LEAVES, EGGPLANT, ESCAROLE, FRISEE, RADICCHIO

FLAVOR IS INFLUENCED BY SEASONALITY, RIPENESS WHEN PRODUCE IS PICKED, EXPOSURE TO THE SUN, WATER, GROWING CONDITIONS, PLANT STRESS, AND THE PYSIOLOGY OF OUR PERSONAL PALATE.

I maintain the reason these folks don't like veggies is because they don't know that spicy ones exist.

What about those of us with timid palates? People who have eaten the same few vegetables all their lives? Depending on the flavor of those few vegetables, look in the vegetable categories that are similar in flavor and start there. That's why I gave examples of sweet and spicy. If you are fearful of all vegetables, start in the mild/neutral category. These vegetables are less assertive in flavor generally and also take on the flavor of the liquids, spices, or dressings you pair with them. No hidden flavor frights usually come with tasting these! Now the flavor family tool is a general guidance tool. You don't have to agree with how I have arranged all the vegetables, and I have only included common vegetables. Use this as a guidance tool and not the definitive guide to vegetable flavors. *And remember, the reason I developed this tool is to guide you to eat more vegetables and common herbs.* Both are rich sources of the many bioactives and nutrients that we learned previous are so positive to healthy genes and gene function.

What About Texture?

Ah yes, I thought about that too! Many people, especially children, are sensitive to texture. I have had many parents look at this tool

Textures – Fruit

SOFT	BANANA, MANGO
SOFT-CHEWY	APPLES (RED, GOLDEN DELICIOUS), APRICOTS, CANTALOUPE, DATES, FIGS, GRAPES, KIWI, MANGO, PEACHES (YELLOW), PEARS (BOSC, COMICE), RASPBERRIES, STRAWBERRIES, WATERMELON
CRISP-CHEWY	GRAPEFRUIT, LEMONS, LIMES, ORANGES
CRISP-CRUNCHY	APPLES (GRANNY SMITH), BLUEBERRIES, CRANBERRIES, HONEYDEW MELON, NECTARINES, PEACHES (WHITE)

Textures – Vegetables

SOFT-CREAMY	AVOCADOS, ARTICHOKES, CELERY ROOT, EDAMAME, GARDEN PEAS, POTATOES
SOFT-CHEWY	ASPARAGUS, BROCCOLI, CARROTS, CAULIFLOWER, EGGPLANT, LEAFY GREENS, LEEKS, MUSHROOMS, ONIONS, PARSNIPS, SWEET POTATOES, TOMATOES (RAW), WINTER SQUASH
CRISP-CHEWY	BEETS, BROCCOLI (RAW), CABBAGE, CARROTS (RAW), ENDIVE, FENNEL, GREEN BEANS, KOHLRABI, LEAFY GREENS (RAW), RADISHES, SNOW PEAS, TURNIPS, ZUCCHINI
CRUNCHY	ASPARAGUS (RAW), BELL PEPPERS (RAW), BOK CHOY, CELERY, ENDIVE (RAW), FENNEL (RAW), RADICCHIO (RAW), SALAD GREENS, SCALLIONS, TURNIPS (RAW)
BUTTERY	AVOCADO, SPINACH, BOSTON BIBB, ARTICHOKE, ETC.
STARCHY	EDAMAME, GARDEN PEAS, POTATOES, ETC.

and identify their kid's eating strategies. Lots of kids (and adults too) don't like soft textures. It's why they may not like peas or mushrooms, avocados, too. It has nothing to do with the flavor of any of these foods. They like crunch. Then don't give them soft food. Hello! Raw vegetables and fruit fit the crunch category, too, so mixing cooked with raw provides you with the texture you are looking for and the nutrients, too.

And Fruit Flavor Families?

Here's the fruit tool. It is organized a little differently, because fruit acceptance mirrors our preference for sweet (less acidity) over tart (more acidity). Texture also plays a role, hence the inclusion of the crisp-fresh category. Again, use this as a general guide to help you make your own discoveries.

Fruit Flavor Families

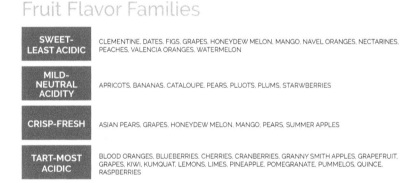

SWEET-LEAST ACIDIC	CLEMENTINE, DATES, FIGS, GRAPES, HONEYDEW MELON, MANGO, NAVEL ORANGES, NECTARINES, PEACHES, VALENCIA ORANGES, WATERMELON
MILD-NEUTRAL ACIDITY	APRICOTS, BANANAS, CATALOUPE, PEARS, PLUOTS, PLUMS, STARWBERRIES
CRISP-FRESH	ASIAN PEARS, GRAPES, HONEYDEW MELON, MANGO, PEARS, SUMMER APPLES
TART-MOST ACIDIC	BLOOD ORANGES, BLUEBERRIES, CHERRIES, CRANBERRIES, GRANNY SMITH APPLES, GRAPEFRUIT, GRAPES, KIWI, KUMQUAT, LEMONS, LIMES, PINEAPPLE, POMEGRANATE, PUMMELOS, QUINCE, RASPBERRIES

After Flavor, What Comes Next?

You have just learned that a wide and receptive fruit and vegetable palate is an important step to optimizing your food choices and subsequently providing the best information for your health. But obviously there's more to eating than fruit and vegetables. The GK Ingredient Toolbox has lots more ingredients beyond this plant category. The question is, what is the best way to organize your kitchen while thinking about how to logically structure meal ideas so that you can get a wide variety of these ingredients, and your other favorites, on your plate? Let me introduce you to my Culinary Framework Tool.

Basic Culinary Framework

PROTEIN
FOCUS: DIVERSE PLANT, SEAFOOD AND ANIMAL PROTEIN OPTIONS

Legumes	Seafood	Animal
DIPS & SPREADS	PATTIES	GRILL
CURRIES & DAHLS	FISH FILET	ROAST
SALADS	SALADS	STIR FRY
		PRESSURE COOK-SLOW COOK

GRAINS
FOCUS: GRAINS IN A VARIETY OF MEAL OPTIONS

Lunch dinner	Breakfast
BOWLS	SMOOTHIES
ROLLS/WRAPS	WARM CEREAL
SALADS	COLD CEREAL
	SAVORY

GREENS
FOCUS: NUTRIENT AND BIOACTIVE POWERHOUSE

Raw

Cooked
SAUTÈ
STEAM

HEALTH SUPPORTIVE SALADS
FOCUS: DELIVERY MEDIUM FOR CRUCIFEROUS VEGETABLES AND NUTRIENT- RICH PRODUCE

CHOPPED & COMPOSED SALADS
SLAWS
SEASONAL FRUIT SALADS

ENERGIZERS & SNACKS
FOCUS: ALTERNATIVES TO STORE-BOUGHT, GRAB AND GO PRODUCTS

Bars

Energy Balls

FLAVORIZERS

Vinaigrettes	Flavored Oils
Pestos	Salsas

Culinary Framework Tool: How to Develop Your Meal Strategy Without Going Nuts

I have already shared a number of strategies to quickly preparing many ingredients in the Genomic Kitchen Ingredient Toolbox and getting them on your plate. My Culinary Framework Tool allows you to take a step back and organize your food thinking into recipe selection and meal preparation ideas without having to dig too hard. I created this tool for nutrition experts whose work focuses in on helping people make the best food choices for their health. Health is a complex issue. In clinical practice, health experts spend a lot of time helping individuals understand their root health concerns and how nutrients impact them. This being the case, the meal planning part of the conversation is often overshadowed by the health explanation part. To ensure meal planning remains in the conversation, I created a series of culinary framework tools to be used as an educational and organizational tool to assist these conversations. Below is my most popular tool. Let's learn how to use it.

Navigating the Tool

The first thing to note is how the tool is organized. Working from the left-hand side, starting at the top, we'll work our way down and across by category. Notice on the left top to bottom we have the six major areas: protein, grain, greens, health supportive salads, energizers, and snacks and flavorizers. As we move to the right across each category, you will find the boxes organized into *suggested* recipe concepts, *which are recipe ideas rather than directives on what specifically to eat.* I have expanded on these concepts in the bulleted sections below, also by category. Some of you may not eat grains or need any form of a snack. This is an organizational tool, not a food directive or suggested intake tool! Let's dig into it, first, then I'll show you how to use it as a guide to choosing and organizing recipes.

PROTEIN

- Protein is organized into plant sources, seafood, and animal protein.

- In the plant and seafood categories, I suggest the types of recipes that can be made using these protein sources. Does anything look familiar? Go back into my kitchen chapters, and you'll see some recipes that fit here.

- In the animal category, I don't suggest recipes, rather I suggest cooking techniques. As you know from me already, those of us who are meat eaters don't need help with meat-based recipes. I leave these recipe categories blank but suggest you vary your cooking techniques, including using fantastic time-savers in the form of a pressure cooker or a slow cooker.

GRAINS

You've probably noticed that I don't include many grains in our GK Ingredient Toolbox. This is *not* because I don't like them. I do! The GK Ingredient Toolbox is built based on current nutrigenomic research and a focus on foods rich in target nutrients and bioactives. In the toolbox, you will find quinoa (technically a seed), wheat germ, and rye. Do not feel restricted to these grains. Remember, my mantra is inclusion not exclusion. That being said, I also want to note that not everyone is gluten-free and not everyone hides from grains. If you are a grain eater, the best grains are the ones in their whole form (hulled or unhulled) that you cook from scratch, many found in the figure below. Whole grains still contain most of their nutrients. Extruded ones, otherwise known as weirdly shaped ones like donut circles or waffle textured squares that that have no resemblance to their field versions, are not what I call "whole forms," regardless of what the Nutrition Facts Panel says. Your body really likes food in its most natural form and not in its extruded, stripped, and reshaped form.

- I subdivided grains into meal parts: breakfast, and lunch/dinner.

- Lunch and Dinner: recognize those recipe strategies? Here I am talking about using the bowl strategy I told you about in Influencers in the Kitchen, Chapter Nine. If you want a bowl-type recipe with a grain base, go to my Go-to Tabouli Salad in the Super Foods Kitchen chapter.

· Breakfast: not everyone is a cereal eater, but for those of you who relish oatmeal, or a warm breakfast of grains with fruit and nut accompaniments, for example, this is where you slot in these recipes. Again, I do not include many grains in the GK Ingredient Toolbox, but this is not because I am suggesting you avoid them. If cereal is your thing, make sure you amp it up with fruit, nuts, and seeds from the GK Ingredient Toolbox and that the base ingredient is whole grain.

Here's my hand-dandy Go-To Guide For Grains for those of you who eat them and love them:

Interchangeable Grains

FAST COOKING < 15 MINS	BUCKWHEAT ,BULGUR, COUSCOUS, QUINOA
MEDIUM COOKING 15-30 MINS	MILLET, WHITE RICE
SLOW COOKING > 30 MINS	BROWN RICE, BARLEY, FARRO, KAMUT, SPELT, RYE/WHEAT BERRIES, WILD RICE
HEARTY	BARLEY, BROWN RICE, FARRO, KAMUT, SPELT, SPECIALTY RICE *WILD FORBIDDEN BHUTANESE, WHEAT OR RYE BERRIES
GLUTEN-FREE**	*BUCKWHEAT, MILLET, *QUINOA RICE
HIGHEST FIBER**	KAMUT, BULGUR, BARLEY, RYE BERRIES
HIGHEST PROTEIN	WILD RICE, RYE BERRIES, MILLET/BUCKWHEAT, BARLEY, BROWN RICE, WHITE RICE
SOFT	BARLEY, COUSCOUS, MILLET, *QUINOA, WHITE RICE

*NOT BOTANICALLY A GRAIN. **FOR GRAINS LISTED HERE.

- Smoothies: no need for me to post a recipe, but I will say, please add a handful of greens and a healthy dose of some form of protein such as nut butter, or a dollop of yogurt. Smoothies without a protein or fat are just a carb juggernaut.

- Savory: this category refers to eggs or other forms of protein. Yes to scrambled eggs or tofu. Yes to the hip avocado toast, with a topping of radishes and dill or a smear of goat cheese. Yes to your go-to breakfast skillet. And yes to beans if they fit in your breakfast repertoire. We navigate savory on the breakfast menu well in America, but refer to your GK Ingredient Toolkit and my Culinary Framework for encouragement. Try some new ideas rather than another trip to the local diner with its nutritionally lopsided breakfast menu.

GREENS

You knew that I would shine a spotlight on this category, didn't you?

- Raw Greens: just go back to Chapter Six, "Master Ingredients in the Kitchen," and look for the sidebar: 13 Ways to Help Crucifers Rock Your Genes—and Your Food (also available in slightly different form on my blog.) Lots of tips there on how to slip these ingredients onto your plate in simple ways. If you need a recipe, try out the Kale, Avocado, and Tart Cherry Salad posted in the chapter.

- Cooked Greens: not everyone loves raw greens all the time, me included. The reason I have this category is to remind you that cooked greens also count as nourishing food. By greens, I am referring not only to the leafy varieties like chard and bok choy, but also cabbage and broccoli. The key to cooked greens is not about the perfect cooking-method, rather what you pair them with. This category is therefore less about how you are cooking them, and more about how you make them palatable. As well, let's remember that the whole Genomic Kitchen concept is more about new ways of thinking about food than an out-and-out cookbook. That said, everyone needs a little help trying new things, so here are a few easy ways to add flavor to cooked greens.

- Sauté for a few minutes in ½ cup chicken or vegetable stock, after the initial cooking. Sprinkle with coarse sea salt before serving. (I got his idea from working with a fantastic southern foods chef in Atlanta. Delicious!)

- Mix 1 tablespoon of cumin and 1 tablespoon of smoked paprika with

1 tablespoon of olive oil. Fold into a large bunch greens at the end of cooking until well incorporated.

- Throw in the zest of 1 lemon and a drizzle of olive oil with a pinch of salt. Toss well.

- Simply drizzle with a little olive oil and dust with a small amount of red pepper flakes.

- Add slivered olives or chopped capers.

- Add 1 or 2 tablespoons of fresh thyme with thinly sliced red onion.

- Serve with a tahini or a nut butter sauce.

- Cook in coconut milk with a little curry powder, too.

- Drizzle with 1 teaspoon sesame oil, 2 tablespoons of soy or tamari and a some grated ginger (optional).

Here's my Getting Greens on Your Plate Tool:

Getting Greens on Your Plate

CLEANING GREENS	BASIC KITCHEN PREP	COOKING BASICS	WHAT GOES WITH A BUNCH OF GREENS
SEPARATE LEAVES FROM STEM OR HEAD.	LEAVES WITH SPINES YOU DISCARD: FOR COLLARDS, KALE, TURNIP GREENS: FOLD LEAF IN HALF AND SLICE OUT STEM.	STEAM: 3 MIN. FOR SPINACH,CHARD, UP TO 10 MIN. FOR KALE, COLLARDS.	DRIZZLE WITH OLIVE OIL, PINCH S & P, 1 TEASPOON LEMON ZEST
FILL CLEAN SINK WITH COLD WATER.	LEAVES WITHOUT RIGID STEMS SUCH AS SPINACH, MUSTARD	SAUTÉ: TO PREFERRED TEXTURE.	DRIZZLE WITH OLIVE OIL, PINCH OF S & P, PINCH OF HOT RED PEPPER FLAKES
SWISH LEAVES AROUND TO REMOVE DIRT/SAND. REPEAT UNTIL WATER IS CLEAR.	GREENS, BEET GREENS, SORREL, LETTUCES: WASH AND COOK AS IS.	BOIL THEN SHOCK: BOIL BRIEFLY IN HOT WATER THEN DRAIN AND DUMP INTO BOWL OF WATER	DRIZZLE WITH OLIVE OIL, ADD CUT UP FRESH OLIVES, ADD TEASPOON LEMON JUICE OR VINEGAR.
DRY IN COLANDER OR SALAD SPINNER.	LEAVES WITH STEMS TO REMOVE AND COOK SEPARATELY: CHARD, BOK CHOY. STRIP STEMS, SLICE INTO 1/2" PIECES AND SAUTE OR ROAST IN OIL.	WITH ICE CUBES. ROAST: TEAR LEAVES UP, TOSS IN OIL, ROAST BRIEFLY AT 400°.	SAUTÉ IN ½ CUP CHICKEN OR VEGGIE STOCK FOR A FEW MINUTES, THEN FINISH WITH DASH OF SEA SALT.
			DRIZZLE WITH 1 TEASPOON SESAME OIL, 1-2 TABLESPOONS TAMARI OR SOY SAUCE AND TOSS WITH GRATED GINGER TO TASTE.
			STIR GREENS COOKED ANY WAY ABOVE INTO ANY COOKED BEAN OR GRAIN RECIPE, LIKE RISOTTO

Health Supportive Salads

Hey, it's a salad so it must be supportive of your health, right? Well, you can layer a bunch of salad bar ingredients on your plate and douse them with a bottled dressing. But, *ahem*, those bacon bits and that mound of processed cheddar may or may not be health supportive, and I can't account for the flavor either. I have talked a lot about salads and slaws in the Master Ingredient Kitchen chapter, pointing out that good ones are excellent "delivery mediums" for the core bioactives and nutrients our genes need to function well.

Simple strategy: make a point of chopping up or grating basic salad ingredients that you can toss together. Do this three times a week, and you will deliver nutrients. You can pre-slice cabbage and bok choy and grate carrots and radishes. Worried about getting enough sulforaphane? Just refresh the mix with a couple of extra radishes or a handful of arugula or watercress. Culinary Genomics does not require a perfect recipe, just the right variety of ingredients to support your genes. It requires understanding which ingredients to choose and why.

Here's my Quick Salad Matrix:

The Basic Chopped Salad
Uses Raw, Sliced, Grated, Chopped Or Shredded Ingredients

Grassy (sliced) GREEN BELL PEPPERS, CUCUMBERS, SPINACH, ZUCCHINI	**Nuts, Seeds, Fruit** ANY	**Hearty Greens (thinly sliced)** CHARD, KALE
Sweet (sliced) SNAP/SNOW PEAS, TOMATOES, RED/YELLOW/ORANGE BELL PEPPERS	**Leafy Greens (simply torn or thin sliced)** **MILD SALAD** LOOSELEAF, BOSTON BIBB, GREEN/RED OAK LEAF, ROMAINE, SPINACH	**Hearty Cruciferae (thinly sliced)** SAVOY, NAPA OR RED CABBAGE
Sweet (grated) BEETS, CARROTS, SWEET POTATOES	**TART** SORREL	
Roots/Tubers (grated or sliced) BEETS, CARROTS, KOHLRABI, RADISHES, SWEET POTATO, TURNIPS	**EARTHY-BITTER** DANDELION GREENS, ENDIVE, RADICCHIO, ESCAROLE, FRISEE	
	SPICY ARUGULA, WATERCRESS	

Yes, we have lots of recipes for salads, slaws, and greens on **our Pinterest boards**. Surf the boards and pin the recipes that appeal.

And don't worry, there are loads of vinaigrette and dressing ideas just ahead in the Flavorizers section.

Seasonal Fruit Salads: it's less about the salad and a whole lot more about the season. Here's my advice. Eat fruit in season. Focus on more acidic fruit such as berries and citrus and less on the higher sugar tropical fruits. Yes, they count, too, but don't build your fruit consumption entirely around pineapple, mangoes and melons.

A Salad for the Orange Citrus Season (November-April generally)

1. Choose 2-3 pieces of assorted citrus: grapefruit, navel oranges, blood oranges, for example.

2. Zest 1 orange and set the zest aside. Peel and thinly slice all the fruit.

3. Layer on a serving dish and scatter the zest evenly over the fruit.

4. Drizzle with a little olive oil.

5. Spritz with 1-2 tablespoons lime juice, to taste.

6. Sprinkle ¼ teaspoon of cayenne or a healthy shake of Aleppo pepper.

7. Scatter 1 -2 tablespoons minced cilantro over the fruit.

8. If you want to add avocado or thinly sliced fennel into the mix, do it. Bon appétit!

Energizers and Snacks

There are many cultures around the world that simply don't eat snacks. The US is not really one of them. We eat while we drive, watch TV, at our desks, while we read texts or emails on our phones. That being said, active people and particularly kids engaged in sports do need fuel. If you or your kids fit in this category, you need recipes that provide the nutrients and fuel. Remember my Apricot Sesame Balls in the Super Foods chapter? These are homemade whole food snack foods. Yes, veggies, nuts, homemade trail mixes, boiled eggs, and fresh fruit (with a smear of nut/seed butter) count too.

Here are a few things to keep in mind when thinking about this category.

- Own the snacks you eat. Bypass the snack aisle, except in a pinch. Ditch the pre-packaged energy category and replace with your own food as much as you can.

- Energy products belong at the gym or the field house and not in your house. They are designed to fuel people with high-energy demand—which is not most of us.

- Eat protein and fat at each meal. Don't eschew the carbs, just make them whole foods: vegetables, fruit, nuts, seeds, legumes. If smoothies are your energy food, remember you need protein and a little fat in each sip.

Flavorizers

This is the category that pulls layers of food together if you are not following a specific recipe. I talked about Flavorizers in the Influencers in the Kitchen chapter. Flavorizers are simple dressings or vinaigrettes. They also include favorites like pestos and even salsa. If you are not a cook, but like the idea of putting ingredients together to form a meal, Flavorizers are a must.

Flavorizers can dress up cooked greens when you add, say, tahini or peanut sauce. They can be tossed with all salad ingredients. See my "Go-To" Mustard Vinaigrette for All. They are also easy to make ahead and keep in your refrigerator.

Below is my quick Flavorizer Matrix to get you started using flavorizer ideas in your kitchen.

Basic Flavorizer Matrix

SIMPLE STYLE: Shake ingredients together in a jar, or whisk together in a bowl.

SIMPLE	RED WINE VINAIGRETTE	SPICY CITRUS HONEY MUSTARD	ASIAN STYLE I
3 PARTS OIL	¼ CUP RED WINE VINEGAR	2 TABLESPOONS ORANGE CITRUS JUICE	2 TABLESPOONS RICE WINE VINEGAR
ONE PART VINEGAR	1 TABLESPOON DIJON MUSTARD	2 TABLESPOONS LIME JUICE	3 TABLESPOONS SOY OR TAMARI
1 TBSP CHOPPED FRESH HERBS	SEA SALT TO TASTE	2 TABLESPOONS OLIVE OIL	1 TABLESPOON HONEY
OR	¾ CUP OLIVE OIL, OR LIGHT FLAVORED OIL, OR EQUAL PARTS OIL AND WATER/STOCK	1 ½ TABLESPOON AGAVE, MAPLE OR HONEY	1 TEASPOON SESAME OIL
4 PARTS OIL			1 TEASPOON TOASTED WHITE SESAME SEEDS (OPTIONAL)
1 PART CITRUS		1 TEASPOON MUSTARD POWDER	
1 TBSP CHOPPED THYME		⅛ TEASPOON CAYENNE	1 TEASPOON GRATED GINGER (OPTIONAL)

CREAMY STYLE: Use a Blender/Small Processor to make these flavorizers

ASIAN STYLE II	NUT/SEED BUTTER	TAHINI	AVOCADO
2 TABLESPOONS SWEET MISO	¼ CUP FRESH SMOOTH NUT OR SEED BUTTER	½ CUP TAHINI	1 AVOCADO PEELED, PITTED
¼ CUP RICE OR WHITE WINE VINEGAR	¼ CUP HOT WATER	½ TEASPOON SESAME OIL	⅓ CUP ORANGE JUICE
2 TABLESPOONS MINCED GINGER	2 CLOVES GARLIC, FINELY MINCED AND MASHED	¼ CUP LEMON JUICE	5 TABLESPOONS LIME OR LEMON JUICE
1 CLOVE GARLIC	2 TABLESPOONS SOY SAUCE	2 TEASPOONS MINCED GARLIC	¼ CUP CHOPPED CILANTRO OR BASIL
¼ CUP LIGHT FLAVOR OIL (OLIVE/CANOLA)	1 TEASPOON GROUND CUMIN	1 TEASPOON RICE VINEGAR OR WHITE/RED WINE VINEGAR	1 CLOVE GARLIC, CHOPPED
	PINCH CAYENNE	¼ CUP WATER (+/- TO ACHIEVE CONSISTENCY)	¼ TEASPOON SALT
	1 TEASPOON FRESH LEMON	PINCH CAYENNE, OR ½ TEASPOON CUMIN OR SMOKED PAPRIKA TO YOUR TASTE	DASH PEPPER OR CAYENNE
			OLIVE OIL OR WATER TO EXTEND DRESSING (OPTIONAL)

Choosing and Organizing Recipes Using the Culinary Framework

Now that you understand how the Culinary Framework is designed, let's talk about how to use it to choose recipes that lead to meals. This is where the internet steps in, and you can leverage free online recipe selection and organization tools. Take a look again at the Culinary Framework which I have added on the opposite page. This time you see that I have added sample recipes from the book to a corresponding category in the framework.

Basic Culinary Framework with Extensions

PROTEIN FOCUS: DIVERSE PLANT, SEAFOOD AND ANIMAL PROTEIN OPTIONS	**Legumes** DIPS & SPREADS: MINTY AVOCADO EDAMAME SPREAD CURRIES & DAHLS SALADS	**Seafood** PATTIES: CURRIED SALMON FISH FILET SALADS: NO FUSS TUNA SALAD	**Animal** GRILL ROAST STIR FRY PRESSURE COOK-SLOW COOK
GRAINS FOCUS: GRAINS IN A VARIETY OF MEAL OPTIONS	**Lunch dinner** BOWLS ROLLS/WRAPS SALADS	**Breakfast** SMOOTHIES WARM CEREAL COLD CEREAL SAVORY	
GREENS FOCUS: NUTRIENT AND BIOACTIVE POWERHOUSE	**Raw** **Cooked** SAUTÉ STEAM		
HEALTH SUPPORTIVE SALADS FOCUS: DELIVERY MEDIUM FOR CRUCIFEROUS VEGETABLES AND NUTRIENT- RICH PRODUCE	**Chopped & Composed Salads** KALE, AVOCADO AND TART CHERRY SALAD **Slaws** ROOT VEGETABLE AND APPLE SLAW	**Seasonal Fruit Salads**	
ENERGIZERS & SNACKS FOCUS: ALTERNATIVES TO STORE-BOUGHT, GRAB AND GO PRODUCTS	**Bars** **Energy Balls** NO-BAKE SESAME APRICOT BALLS		
FLAVORIZERS	**Vinaigrettes** TAHINI, ASIAN STYLE **Pestos**	**Flavored Oils** **Salsas**	

Basic Culinary Framework

WHERE TO START

The Culinary Framework is a guide to thinking about how you can slot in some of your own favorite recipes, and maybe adapt some of them using the tips and strategies I have shared in the book. Simple nips and tucks using ingredients from the Genomic Kitchen Ingredient Toolbox allow you to take your recipes to a new level of positive support for your genes and your health. If you are a little low on recipes and need some inspiration, here are two strategies for you to consider.

1. Use Pinterest! You can start by visiting the **Genomic Kitchen Pinterest boards**, which I have introduced you to already. There you will find a wide variety of recipes chosen specifically for the ingredients they contain or the culinary category into which they fit. For example, you'll find flavorizers on the Dips, Dressings and Spreads board. You can find a variety of recipes using legumes on the board with the same name. The same applies to salads and slaws. Or you can search for recipes by a specific bioactive, for example quercetin or lycopene. Or, search for recipes by a specific MISE category. These are labeled Genomic Kitchen Influencers (I), or Genomic Kitchen Super Foods (S).

2. Download the app **Plan to Eat** or the **Paprika Recipe Manager** app. Both also work on your desktop as well. I have been using these apps (installed on my laptop) for a quite a while. You can import a recipe you like from any culinary or lifestyle magazine directly into your account and tag it any way you like. Specific to Plan to Eat, in the same way that you can pin a recipe to your Pinterest boards, you can download a "Plan to Eat" (PTE) icon which sits in your browser tabs. When you see a recipe you like, you click on the PTE icon which automatically transfers the recipe information and recipe image to your account. You can also manually add your own recipes. Paprika does not have a browser extension "pin," however you can navigate to any recipe using Paprika's browser option and then click "save recipe" at the bottom of the screen and automatically import the recipe to your Paprika files. Paprika also allows you to import and save "how to" videos which is extremely useful. I love these highly flexible apps which make choosing, organizing and finding recipes so easy.

3. **Eat Your Books** is another very useful app that helps you search for recipes in cookbooks you own. You simply search for and add the cookbooks you own into an online library. The app then classifies all the recipes in those cookbooks into a searchable format that allows you to screen by style of cuisine, type of ingredient, or even

recipes with an online video option. You therefore have a quick online searchable snapshot of all your cookbooks allowing you to navigate your collection in an efficient and rapid manner and get to recipes that include the ingredients you are targeting. What a great way to look in your pantry or refrigerator, figure out the ingredients on hand, and then find a recipe in your own cookbook collection that uses what you have available. You still need to refer to the copy of a cookbook you own, be it in print or digital format. However, you can get there fast without hours of flipping. The first five books you include in your library are free. Thereafter you may a nominal annual fee to include as many of your cookbooks as you want.

Fine-tuning Your Own Recipes for the Best Food-Gene Relationship

Hopefully by now you have a good grasp on how to choose and prepare ingredients that influence your genes positively. More importantly, you now understand why we emphasize some ingredients over others. What remains in the Genomic Kitchen is to help you look at your own recipe favorites, or maybe the menu at your local restaurant, and apply what I call a Quality Assurance Technique to see if that recipe or dish pushes the right genetic buttons. If it does not, what substitutions or additions can finesse the creation you are about to prepare or order? Here is my Recipe Quality Assurance Tool.

Recipe Quality Assurance Tool

BLANK LANDSCAPE	DEEP NUTRITION	ADD FLAVOR	BOOSTER
MINIMAL FLAVOR INGREDIENT	SEEK OUT THE CRUCIFERS FIRST	SALSAS	HERBS & SPICES
PROTEIN	OR AT LEAST ONE VEGGIE FROM THE INGREDIENT TOOLBOX	VINAIGRETTES/DRESSINGS	NUTS & SEEDS
LEGUME		PESTO	OLIVE OIL, NUT OIL, FLAXSEED OIL, SESAME OIL, TAMARI SOY,
WHOLE GRAIN	BALANCE LOW/HIGH GLYCEMIC FRUIT AND VEGETABLES	SIMPLE VEGETABLE OR FRUIT BASED SAUCES/CREAMS	LIQUID AMINOS ETC., MISO
			SEA VEGETABLES
			FERMENTED/CULTURED FOODS

Recipe Quality Assurance Tool

Looking at the tool, you can see there are four parts to it.

- **Blank Landscape** refers to either the protein element or the ingredient that adds bulk in your recipe or dish. Notice that I refer to the blank landscape as the "minimal flavor ingredient." For the most part, legumes and grains don't have a robust flavor profile, so we use other ingredients to enhance their flavor. They provide nutrients and calories, but not unique flavor. Animal and seafood proteins do provide flavor, but depending on their freshness and choice of protein, we will add additional elements to enhance flavor. So think of the blank landscape as an integral part of your dish/recipe, but not the focus.

- **Deep Nutrition** refers to the incorporation of vegetables or fruit from the GK Ingredient Toolbox. Here it is important to focus on including at least *one*, preferably many, of the fruit or vegetables I include. I don't need to explain why anymore. Seek out crucifers whenever you can. If they are not available, then choose another ingredient—again, you know why. Another piece of advice when selecting your produce for recipes or at the restaurant is to balance your starchy choices with the non-starchy. Some fruit, like mango and watermelon, is higher in sugar than others. Root vegetables, while nutritionally robust, have a higher starch (sugar) content than leafy vegetables, for example. Higher starch/sugar produce items are referred to as "high glycemic" because their carbohydrate content can raise blood sugars faster than other choices. This may overwhelm your body's ability to bring blood sugar levels back to normal. My advice? Just balance it out. All produce items in the GK Ingredient Toolbox are there for a reason. Look for balance.

- **Add Flavor.** Now this category you know well. What you want to look for in recipes and dishes is how the flavor is added. If this is a dressing, do you know what is in the dressing? Is one dressing, pesto, or salsa more nutritionally robust than other? Choose an herbed dressing over a plain one, for example. Choose a salsa featuring one or more ingredients from the Genomic Kitchen Ingredient Toolbox. Can you choose a sauce made from puréed vegetables or fruit from the Genomic Kitchen Ingredient Toolbox? Oh—creams refers to creamy sauces made from yogurt for example, or avocado or nuts. Think of guacamole as a "cream." A satay sauce, rich and peanut-based, could be considered a creamy Flavorizer, for example. In short, you are looking for those nips and tucks that bump up the nutritional aspects that benefit your genes.

- **Booster**: Not every recipe or every dish will have a target ingredient from each MISE category. What if you have a recipe that has only one ingredient? The Booster category addresses this deficit. Depending on the booster category you choose, you are adding:

- **A Master Ingredient nutrigenomic boost** through herbs and spices.

- **An Influencer Ingredient boost** through olive oil or nut oils.

- **A Super Food**, or super nutrient boost, through nuts, seeds or sea vegetables

- **An Enabler boost** through fermented or cultured foods such as yogurt, aged cheese (go lightly here), or the addition of naturally fermented pickles. Recipes or menu items that feature tamari, soy, miso, or fish sauce, for example, are also ways to integrate ingredients that support you gut. But let's face it, in commercial food establishments the chances are that many of these products may be mass-produced and therefore not a rich source of probiotics. You might want to ask your server or chef.

Here's an Example from a Restaurant

Let's say you are on the road and pull into a place that has a salad bar, or maybe a salad option. Instead of randomly piling on ingredients to fill out your plate, think more strategically.

- Start with the protein that compels you. Maybe it is those ubiquitous kidney beans (that always require some flavor), or perhaps there is some shredded chicken. There's your foundation.

- Next, you're after the deep nutrition to be supplied by vegetables. Bypass the shredded pale-looking lettuce and head for the kale, salad mix, or spinach. Pile it on because you know that sulforaphane is going to boost your gene function! Yes, you can add shredded cabbage and certainly some of the beets, definitely some of the broccoli or cauliflower that is usually lurking there.

- Then we get to the flavor part. Often this is displayed in squeeze bottles or wells filled with commercially prepared salad dressing consisting of ingredients of unknown origin. Look beyond these to the bottles of olive oil and vinegar. This is where you add flavor. If you want, use the oil and ask for half a lemon or lime to add your own acidity. If you stick with the vinegar, know that balsamic will add fruitiness and some sweetness if your palate tends that way. But, please, leave the pre-formulated honey-Dijon or raspberry vinaigrette alone.

- Finally, for that nutrient booster, look for nuts and seeds and sprinkle them on top.

- Now return to your table, mix everything together and you have created (and controlled) a meal whose components will now nourish your genes.

Let's use another example, this time using a visit to the Panera Bread restaurant chain. Here you can order directly at the counter, or navigate through their computer-driven ordering system which includes salads and bowls. The computer navigation menu allows you to look at the ingredients for each dish so that you can choose with MISE in mind. At the time of this writing, Panera offers an Ancient Grain, Arugula and Chicken Salad reflecting ingredients in our M, I, and E categories. Their Soba Noodle Bowl with Edamame provides ingredients from each of our MISE categories where cabbage provides M, mushrooms and spinach provide I, sesame seeds and edamame provide the S, and the miso broth provides the E. Wow! And, if you are still hungry, or want to go in a different direction, you can find a Greek Yogurt parfait containing ingredient options from both M and E.

Here's the thing. When you eat out, use the MISE strategy (the ingredient categories that comprise the Genomic Kitchen Ingredient Toolbox) to guide your choices, but don't drive yourself nuts trying to create meals that include ingredients from every category. Strive for two or three categories. If you hit four categories then congratulations! MISE provides you with an understanding of how specific ingredients function with your body, biochemistry and health, but it's not a standing order. Strive to include a mixture of these ingredients every single day, remembering that life and eating is about balance, not an ultimatum.

My Recipe Quality Assurance Tool is designed to help you think through how to choose dishes or adapt recipes to maximize their benefit to you and your health. Use it if it helps you. If I have done my work correctly for you and unpacked MISE and the Genomic Kitchen Ingredient Toolbox in a way that resonates for you, then you are all ready to hit the kitchen and start cooking to communicate in the language of your DNA.

Where to from Here?

I started this book journey with a visit to Ristorante AGA in Northern Italy. If you recall, I ate a meal that was a revelation for me. The revelation was in the creative use of ingredients that connected to my genes through their inherent properties (bioactives), and the way that they were prepared. I was eating food that "talked" to my genes! Granted, my background in nutrigenomics allowed me to connect the dots, but it also told me very clearly that this was not a conversation I wanted to keep to myself, rather one that I wanted to share with *you*. Wouldn't it be marvelous if we could all hop a flight and sit down at a quiet restaurant in Northern Italy and indulge in a culinary marvel? I know I would go back again tomorrow.

Since that is not an everyday possibility for most of us, what we can do is to take charge of our health through the food that we choose. Whether you are at the grocery store, in your kitchen, browsing through cookbooks or Pinterest, or even dining at your local or destination restaurant, it is my hope that the lens through which you view your food choices and preparation has changed.

Nutrigenomics is redefining how we understand what food does in our body. Food is no longer viewed as a parcel of nutrients wrapped in a recipe that is simply "good for us." It is also not predestined for a particular body part, such as your bones or your heart, simply because it contains calcium or vitamin D.

Food is a communication tool, where receptors in your genes are the chief receivers and users of that information. For sure, the most personalized insights come from a genomic test that reveals how *your* unique blueprint of genes responds to food and the environment you live in. But I did not write this book to advocate for personalized genomic testing, rather as an introduction to a new way of thinking about and preparing food that facilitates the most useful relationship with everyone's genes. Human genes are programmed to respond to food and utilize it to support the biochemistry that defines our health. While differences between us define how much of a specific food or nutrient we each need, the fact remains that food works essentially in the same way for all of us.

In this book I have provided you with a toolbox whose ingredients will shower your genes with the information they need to better support your health. I have explained why we focus on these ingredients and how they work. But remember, these ingredients are a guide for you, a foundation on which you can build your pantry and refrigerator. You do not need to exclude foods that you love or recipes your family can't live without. Mostly, I want you to know the relevance of each ingredient you spend money on and then to make food choices and meals that both nourish you and soothe you.

Let me leave you with this one final note. Food is the information source that feeds and fuels your body. Food is also a means of bringing human beings together. Let us not forget that sharing a meal with others is also feeding our genes. You see, food can never replace the ambience or love of eating together. Our genes respond to love and togetherness and the laughter that ensues from gathering around the table. The table is a place where people can come together to share, laugh, cry, and touch each other. Connectivity as humans is as important as the bread we break and the food we share. Rejoice with good food, lots of laughter, and the power of human embrace and love.

Appendix A: Genomic Testing. For Me or Not for Me?

We are entering an era of personalized medicine. This means that we move from applying the same therapy to you as your neighbor (assuming you have the same problem!), to a therapy that is adapted to you as a unique individual.

Consider this statement from a 2011 paper, The n-of-1 clinical trial: the ultimate strategy for individualizing medicine: "There is a growing acceptance that the development of medical interventions that work ubiquitously (or under most circumstances) for the majority of common chronic conditions is exceptionally difficult and all too often has proved to be fruitless." All you have to do is read the pages of disclaimers that follow a magazine pharmaceutical ad to know that any one medication may work, but not for everyone and not under all circumstances. So how do *you* know? Unless you have genomic information that indicates how well you metabolize or handle a particular drug, you don't. You find out the hard way when that medication makes you feel worse! This form of genomic testing is called pharmacogenomics and provides clarity around how your genes might handle different classes of drugs.

So why would you want or need genomic testing? Genomic

testing provides the clearest insights into your gene operating system. You already know about SNPs (single nucleotide polymorphisms) and how they mirror your unique blueprint which is different from anyone else's. Genomic tests provide information about how your genes are functioning. For example, a genomic test could provide useful insights into how your body manages different fats and carbohydrates. Armed with this information, you may be advised by your healthcare provider to alter the types of fat you are eating and the amount of carbohydrate, too.

Let's say you eat a diet rich in fruit and vegetables and you do a hard workout at the gym four days a week yet you struggle with your weight. How come you can't get the weight off, but others can? Genomic information can pinpoint which genes might be sluggish in carbohydrate or fat metabolism, or which gene could affect how your body turns on and off your satiety (appetite) mechanism. Armed with this information, we might tailor exercise and diet differently to optimize gene behavior (expression) and help those genes do their job a little more efficiently, subsequently helping you.

Breast cancer is a prevalent form of cancer. Estrogen is often pointed to as a culprit, among other factors. Genomic testing provides insights into how your body is handling the process of detoxifying, or breaking down estrogen, into less harmful compounds to be removed from the body. Left to its own devices, estrogen can cause havoc in the body if it is not properly disposed of. Skilled clinicians use a polygenic approach, meaning they look at multiple genes in the estrogen metabolism pathway, to compile a picture of how your body metabolizes estrogen. If your genes line up with minimal SNPs, you are probably metabolizing and disposing of estrogen in a harmless way. But what happens if testing shows that you have multiple SNPs in the estrogen pathway? The good news is that genomic information provides the right information for clinicians to order laboratory tests to assess biomarkers (biological indicators) that determine where your estrogen biochemistry is (or is not) getting hung up and how we can intervene. Again, trained clinicians know how to interpret the genomic test, order appropriate labs, assess those results, and then develop a nutrition and lifestyle protocol with you to get your health on the right track, and give your genes a helping hand too.

Genomic testing is powerful, representing the cutting edge of nutrition science and personalized health management. The question is: who is a good candidate for genomic testing, or, otherwise put, who could benefit the most from genomic testing and is that person *you*? Working with clinicians practicing genomic medicine, a number of answers emerge. Experienced clinicians working with genomic information in clinical practice say that without it, we are often looking for a needle in a haystack when trying to resolve persistent health issues. I happen to agree.

So the bottom line is this. If you are a person living with persistent unresolved health issues, finding yourself moving from doctor to doctor and getting no answers, then yes, genomic testing, provided by a trained clinician is for you. If you are motivated about your health and want to apply the next level of personalization to your healthcare, then genomic testing is also for you. And as you now know, if you are happy with your health, don't want to pay for genomic testing, but want to fine-tune it a little more by applying principles of nutrigenomics to your pantry and your plate, then follow the principles I have laid out in this book. You can start to take advantage of the science of nutrigenomics and get the right food on your plate.

Appendix B: SIRT-1 and PPARG Appendix

SIRT-1: A Simplified Overview of How This "Ringmaster" Gene Manages Metabolism

SIRT-1 is a ringmaster, playing an interactive role with a number of important genes that influence how your body manages inflammation and blood sugars and determines whether it burns or stores fat. SIRT-1 is one of a seven protein member family called SIRTUIN, and it is the most studied and understood of this complex family because of its importance.

One of the genes that SIRT-1 interacts with is a metabolic giant called AMPK (AMP-Activated Protein Kinase). When energy levels drop in the cell, AMPK steps into action to generate energy, choosing fat as its main fuel to burn, thus accessing the body's fat deposits. (It could also burn protein, but generally it is not a good thing for the body to be generating energy by diminishing muscle mass, for example.) Interestingly, the drug metformin, used in the treatment of Type II diabetes, targets this particular gene for activation, helping reduce the impact of fat accumulation in the body, obesity being one of the hallmarks of kind of diabetes. AMPK also helps reduce oxidative stress in the body, which we know can be a root cause of disease. AMPK is therefore a very important and useful gene.

SIRT-1 can also sense when energy levels drop either through exercise or the fact that you need to eat, and can activate AMPK. AMPK then steps up to the task of energy production by accessing your fat deposits to produce energy. Once energy levels are restored, AMPK turns off SIRT-1, stopping the fat burning process. SIRT-1 and AMPK thus have a two-way relationship that benefits you and your metabolism. Note that too many calories (overeating) and high levels of blood sugar can actually deactivate both AMPK and SIRT-1, thus eliminating the favorable burning of available body fat as an energy source. Fat as a fuel source is a good thing, and the SIRT-1 and AMPK partnership enables you to preferentially burn fat and reduce those fat stores.

We know that SIRT-1 and AMPK interact synergistically with each other, but what if we could drive the partnership using food, essentially stimulating SIRT-1 to activate AMPK? This is where resveratrol comes in because it can activate SIRT-1 in turn activating AMPK. We know that AMPK can be independently activated by a drop in energy, but if we use SIRT-1 to add fuel to the metabolic fire through what we eat, we can add a further stimulus to the energy production/fat burning process.

NOTE: AMPK can be activated independently by a drop in energy, but in tandem with SIRT-1 we have a dynamic process for stimulating energy production. Research appears to validate the resveratrol activates SIRT-1 which then activates AMPK. Resveratrol does not appear to upregulate AMPK directly.

PPARG and Metabolism

Metabolic gene giant number two is PPARG, or Peroxisome proliferator-activated receptor gamma. PPARG is a member of an important family of genes instrumental in metabolism. PPARG, specifically, plays an essential role in how your body handles fat and blood sugar. When PPARG is activated, it stimulates fat storage or fat accumulation in your cells. As such, this gene has been associated with numerous diseases including diabetes, atherosclerosis, and even cancer. The way things work in the body, if you are storing fat, you can't preferentially burn it for energy. Both mechanisms do not work well together. And if you want to maintain your weight (and your health), you want to burn fat and not accumulate it! Enter SIRT-1 again. SIRT-1 has the ability to repress PPARG, essentially slowing the process of fat accumulation, making fat available as a preferential fuel. Concurrently, repressing PPARG also appears to increase insulin sensitivity, thus creating a perfect scenario for managing both blood sugars and lipids. Interestingly, thiazolidinediones are a class of "agonist" drugs that target PPARG and are commonly used to treat Type II diabetes. By blocking the activity of PPARG, insulin sensitivity is improved in muscle, liver and fat cells. Insulin is a hormone essential to blood sugar management. Diabetes is a disease that requires medical supervision, so I do not suggest you substitute red wine and grapes (sources of resveratrol) for your current medications. But now you understand how SIRT-1 interacts with a gene that can impact how you manage fat and blood sugars, and so choose to include one or both of these foods in your diet.

References

Introduction

The Genetics of Exceptional Longevity:

Santos-Lozano A , Santamarina A, Pareja-Galeano H, et al. The genetics of exceptional longevity: Insights from centenarians. *Maturitas*. 2016. 90:49-57.

Individualized Medicine:

Lillie EO, Patay B, Diamant J, Issell B, Topol EJ, Schork NJ. The n-of-1 clinical trial: the ultimate strategy for individualizing medicine? *Per Med*. 2011;8(2):161-173.

SNPs and ACE gene and blood pressure:

Uh ST, Kim TH, Shim EY, Jang AS, Park SW, Park JS, et al. Angiotensin-converting enzyme (ACE) gene polymorphisms are associated with idiopathic pulmonary fibrosis. *Lung*. 2013. 191(4):345-51.

Genes and Longevity General:

Shivani Garg, MD, MBBS; Chief Editor: Karl S Roth, MD. Alzheimer Disease and APOE-4. Medscape. https://emedicine.medscape.com/article/1787482-overview. Updated February 1, 2015. Accessed November 27, 2018.

Prada D, Colicino E, Power MC, et al. Influence of multiple APOE genetic variants on cognitive function in a cohort of older men - results from the Normative Aging Study. *BMC Psychiatry*. 2014;14:223.

F.A. Sayed-Tabatabaei, B.A. Oostra, A. Isaacs, C.M. van Duijn, and J.C.M. Witteman. ACE polymorphisms. *Circ Res.* 2006;98:1123–1133.

Chapter One

AGA and The New York Times:

Draper, Robert. In Italy, Hiking and Haute Cuisine in the Dolomites. *The New York Times.* Aug. 16, 2015. https://www.nytimes.com/2015/08/16/travel/italy-dolomites-hiking-trekking-camping.html. Accessed Feb 22, 2018.

Genomic Medicine:

What is Genomic Medicine? National Human Genome Research Institute. https://www.genome.gov/27552451/what-is-genomic-medicine/. *Updated: November 6, 2018. Accessed November 1, 2018.*

Green Tea:

Chacko SM, Thambi PT, Kuttan R, Nishigaki I. Beneficial effects of green tea: a literature review. *Chin Med.* 2010;5:13.

Khan N, Mukhtar H. Tea and health: studies in humans. *Curr Pharm Des.* 2013;19(34):6141-7.

Chapter Two

Bioactives in Foraged Greens, University of Athens study:

Trichopoulou A, Vasilopoulou E, Hollman P, Chamalides C, Foufa E, Kaloudis T, et al. Nutritional composition and flavonoid content of edible wild greens and green pies a potential rich source of antioxidant nutrients in the Mediterranean diet. *Food Chem.* 2000;70:319-323.

Bioactives in diet from herbs and greens:

Kaloteraki C, Velivasaki, M, Tsourdalaki, E. Comparison of wild greens and herbs consumption between residents of urban and rural areas of Crete. *Clinical Nutrition.* ESPEN. Volume 13, e61.

Buettner's NYTimes article:

Buettner, Dan. The Island Where People Forgot to Die. *The New York Times Magazine.* Oct 28, 2012. https://www.nytimes.com/2012/10/28/magazine/the-island-where-people-forget-to-die.html. Accessed May 29, 2018.

Olive Oil and Mediterranean Diet:

Castañer O, Corella D, Covas MI, Sorlí JV, Subirana I, Flores-Mateo G, et al. In vivo transcriptomic profile after a Mediterranean diet in high-cardiovascular risk patients: a randomized controlled trial. *Am J Clin Nutr.* 2013. 98;3:845-53.

Longevity:

Govindaraju D, Atzmon G, Barzilai N. Genetics, lifestyle and longevity: Lessons from centenarians. *Appl Transl Genom.* 2015;4:23-32.

Predimed Mediterranean Diet Study:

https://www.ncbi.nlm.nih.gov/pubmed/25316904

Predimed Study Analysis by Harvard School of Public Health:

Harvard School of Public Health. PREDIMED Study Retraction and Republication. June 22, 2018. https://www.hsph.harvard.edu/nutrition-source/2018/06/22/predimed-retraction-republication/ Accessed Aug 1, 2018.

Impact of Mediterranean diet on cancer: focused literature review:

Barak Y, Fridman D. Impact of Mediterranean Diet on Cancer: Focused Literature Review. 2017. *Cancer Genom Proteom.* 14(6): 403-40.

Polyphenols, Mediterranean diet, and inflammation:

Billingsley HE, Carbone S. The antioxidant potential of the Mediterranean diet in patients at high cardiovascular risk: an in-depth review of the PREDIMED. *Nutr Diabetes.* 2018;8(1):13.

Fermented dairy and Mediterranean diet:

Mena-Sánchez G, Babio N, Martínez-González MÁ, Corella D, Schröder H, Vioque J, et al. Fermented dairy products, diet quality, and cardio–metabolic profile of a Mediterranean cohort at high cardiovas-cular risk. 2018. *Nutr Metab Cardiovasc Dis.* 28(10):1002-1011.

Mediterranean Diet and health and longevity:

Martinez-Gonzalez MA, Martin-Calvo N. Mediterranean diet and life expectancy; beyond olive oil, fruits, and vegetables. *Curr Opin Clin Nutr Metab Care.* 2016;19(6):401-407.

Asian diets and consumption of bioactives:

Imamura F, Micha, R, Khatibzadeh, S, Fahimi, S, Shi, P, Powles, J, et al. Dietary quality among men and women in 187 countries in 1990 and 2010: a systematic assessment. *Lancet Glob Health.* 2015;3(3):PE132-E142.

National Geographic article: "What the World Eats:"

What the World Eats. National Geographic. https://www.nationalgeographic.com/what-the-world-eats/ Accessed date: October 4, 2018.

Human Social Genomics:

Cole, S. Human social genomics. *PLoS Genet.* 2014;10(8):e1004601.

Milk thistle:

Weidmann, A. Dihydroquercetin: more than just an impurity? *Eur J Pharmacol.* 2012;684(1-3): 19-26.

Chronic inflammation and longevity:

https://www.news-medical.net/news/20181102/Chronic-inflammation-linked-with-increased-risk-of-multiple-diseases-and-shorter-lifespan.aspx

Curcumin and obesity and diabetes:

Siriwardhana N, Kalupahana NS, Cekanova M, LeMieux M, Greer B, Moustaid-Moussa N. Modulation of adipose tissue inflammation by bioactive food compounds. *J Nutr Biochem.* 2013. ;24(4):613-23.

Chapter Three

General background on Genomic Medicine:

National Human Genome Research Institute. Genomic Medicine and Health Care. www.genome.gov. https://www.genome.gov/27527652/genomic-medicine-and-health-care/.Last Updated: July 21, 2016. Accessed May 22, 2018.

Mead M. Nutrigenomics: the genome--food interface. *Environ Health Perspect.* 2007;115(12):A582-9.

SNPs and their effect on gene function:

Shastry BS. SNPs: impact on gene function and phenotype. *Methods Mol Biol.* 2009;578:3-22.

National Human Genome Research Institute. FAQ About Genetic and Genomic Science. https://www.genome.gov/19016904/faq-about-genetic-and-genomic-science/. Updated: Sept 7, 2018. Accessed September 29, 2018.

Plant bioactives:

Fraga CG, Oteiza PL, Galleano M. Plant bioactives and redox signaling: (-)-Epicatechin as a paradigm. *Mol Aspects Med.* 2018;61:31-40.

Milner JA. Molecular targets for bioactive food components. *J Nutr.* 2004;134(9):2492S-2498S.

Elsamanoudy AZ, Mohamed Neamat-Allah MA, Hisham Mohammad FA, Hassanien M, Nada HA. The role of nutrition related genes and nutrigenetics in understanding the pathogenesis of cancer. *J Microsc Ultrastruct.* 2016;4(3):115-122.

Quercetin:

Anand David AV, Arulmoli R, Parasuraman S. Overviews of Biological Importance of Quercetin: A Bioactive Flavonoid. *Pharmacogn Rev.* 2016;10(20):84-89.

Mostafavi-Pour Z, Ramezani F, Keshavarzi F, Samadi N. The role of quercetin and vitamin C in Nrf2-dependent oxidative stress production in breast cancer cells. *Oncol Lett.* 2017 Mar;13(3):1965-1973.

Traditional food knowledge, Italy:

Sansanelli S, Tassoni A. Wild food plants traditionally consumed in the area of Bologna (Emilia Romagna region, Italy). *J Ethnobiol Ethnomed.* 2014;10:69.

Traditional Knowledge, Turkey:

Tan A, Adanacioglu, N, Karabak S, Aykas, L, Lerzan, Tas, N, Taylan, T. Biodiversity for food and nutrition: edible wild plant species of Aegean region of Turkey. *J of AARI.* 2017;27(2): 1-8.

Dogan, Y. Traditionally used wild edible greens in the Aegean region of Turkey. *Acta Soc Bot Pol.* 2012;81(4): 329-342.

Lycopene:

Kessy N, Honest H, Wei Z, Lianfu Z. Lycopene: Isomerization Effects on Bioavailability and Bioactivity Properties. *Food Rev Int.* 2011; 27:3, 248-258.

Chapter Four

Mercury poisoning and selenium:

Ministry of the Environment. National Institute for Minamata Disease. http://nimd.env.go.jp/english/index.html. Updated May 25, 2018. Accessed October 29, 2018.

Minamata Bay. https://en.wikipedia.org/wiki/Minamata_Bay. Updated March 21, 2018. Accessed May 3, 2018.

Harada M. Minamata disease: methylmercury poisoning in Japan caused by environmental pollution. *Crit Rev Toxicol.* 1995;25(1):1-24.

Ralston NV, Raymond LJ. Dietary selenium's protective effects against methylmercury toxicity. *Toxicology.* 2010 Nov 28;278(1):112-23.

Bose-O'Reilly S, McCarty KM, Steckling N, Lettmeier B. Mercury exposure and children's health. *Curr Probl Pediatr Adolesc Health Care.* 2010;40(8):186-215.

Kehrig HA, Seixas TG, Di Beneditto AP, Malm O. Selenium and mercury in widely consumed seafood from South Atlantic Ocean. *Ecotoxicol Environ Saf.* 2013;93:156-62.

Glutathione:

Jozefczak M, Remans T, Vangronsveld J, Cuypers A. Glutathione is a key player in metal-induced oxidative stress defenses. *Int J Mol Sci.* 2012;13(3):3145-75.

Burger J, Jeitner C, Donio M, Pittfield T, Gochfeld M. Mercury and selenium levels, and selenium:mercury molar ratios of brain, muscle and other tissues in bluefish (Pomatomus saltatrix) from New Jersey, USA. *Sci Total Environ.* 2013;443:278-86.

Green Tea and Effect on Genes:

Kaul D, Sikand K, Shukla AR. Effect of green tea polyphenols on the genes with atherosclerotic potential. *Phytother Res.* 2004 Feb;18(2):177-9.

Turmeric:

Kim JH, Gupta SC, Park B, Yadav VR, Aggarwal BB. Turmeric (Curcuma longa) inhibits inflammatory nuclear factor (NF)-κB and NF-κB-regulated gene products and induces death receptors leading to suppressed proliferation, induced chemosensitization, and suppressed osteoclastogenesis. *Mol Nutr Food Res.* 2012;6(3): 454-65.

Hot Pepper or Capcaisin:

McCarty MF, DiNicolantonio JJ, O'Keefe JH. Capsaicin may have important potential for promoting vascular and metabolic health. *Open Heart.* 2015;2(1):e000262.

Ginger:

Ho SC, Chang KS, Lin CC. Anti-neuroinflammatory capacity of fresh ginger is attributed mainly to 10-gingerol. *Food Chem.* 2013;141(3): 3183-3191.

Mashhadi NS, Ghiasvand R, Askari G, Hariri M, Darvishi L, Mofid MR. Anti-oxidative and anti-inflammatory effects of ginger in health and physical activity: review of current evidence. *Int J Prev Med.* 2013;4(Suppl 1):S36-42.

Brendan Brazier:

Brazier, Brendan. The thrive diet: the whole food way to lose weight, reduce stress, and stay healthy for life. Da Capo Books: Philadephia, PA. 2008. 309pp.

Honey:

Pichichero E, Cicconi R, Mattei M, Muzi MG, Canini A. Acacia honey and chrysin reduce proliferation of melanoma cells through alterations in cell cycle progression. 2010. *Int J Oncol.* 2010;7(4):973-81.

Mijanur Rahman M, Gan SH, Khalil MI. Neurological effects of honey: current and future prospects. *Evid Based Complement Alternat Med.* 2014;2014:958721.

Bahadori M, Baharara J, Amini E. Anticancer Properties of Chrysin on Colon Cancer Cells, *In vitro* and *In vivo* with Modulation of Caspase-3, -9, Bax and Sall4. *Iran J Biotechnol.* 2016;14(3):177-184.

Luteolin:

Luo Y, Shang P, Li D. Luteolin: A Flavonoid that Has Multiple Cardio-Protective Effects and Its Molecular Mechanisms. *Front Pharmacol.* 2017;8:692.

Wang G, Li W, Lu X, Bao P, Zhao X. Luteolin ameliorates cardiac failure in type I diabetic cardiomyopathy. *J Diabetes Complications.* 2012;Volume 26(4): 259-265.

Olive oil:

Truth in olive oil. www.extravirginity.com

Milestone Olive Oil. http://www.ofdreamsandknowledge.gr/en/

Olea Estates. https://oleaestates.com/

Pruneti. http://www.pruneti.it/en/

Chapter Five

Nrf2 and gene transcription:

Transcription factors. Khanacademy.org. khanacademy.org/science/biology/gene-regulation/gene-regulation-in-eukaryotes/a/eukaryotic-transcription-factors.

SIRT-1:

Ross D, Siegel D. Functions of NQO1 in Cellular Protection and CoQ10Metabolism and its Potential Role as a Redox Sensitive Molecular Switch. *Front Physiol.* 2017; 8:595.

Xiaoling Li. SIRT1 and energy metabolism. *Acta Biochim Biophys Sin (Shanghai).* 2013;45(1):51-60.

Schug TT, Li X. Sirtuin 1 in lipid metabolism and obesity. *Ann Med.* 2011;43(3):198-211.

de Kreutzenberg SV, Ceolotto G, Papparella I, et al. Downregulation of the longevity-associated protein sirtuin 1 in insulin resistance and metabolic syndrome: potential biochemical mechanisms. *Diabetes.* 2010;59(4):1006-15.

SIRT 1 and weight gain/loss:

Zillikens MC, van Meurs JB, Rivadeneira F, et al. SIRT1 genetic variation is related to BMI and risk of obesity. *Diabetes.* 2009;58(12):2828-34.

Weight and body clock and the CLOCK gene:

Maury E, Ramsey KM, Bass J. Circadian rhythms and metabolic syndrome: from experimental genetics to human disease. *Circ Res.* 2010;106(3):447-62.

James SM, Honn KA, Gaddameedhi S, Van Dongen HPA. Shift work: Disrupted Circadian rhythms and sleep-implications for health and well-being. *Curr Sleep Med Rep.* 2017;3(2):104-112.

Zanquetta M, Lúcia Corrêa-Giannella M, Beatriz Monteiro M, Villares, S. Body weight, metabolism and clock genes. *Diabetology & Metabolic Syndrome.* 2010; 2:53.

Chang HC, Guarente L. SIRT1 mediates central circadian control in the SCN by a mechanism that decays with aging. *Cell.* 2013;153(7):1448-60.

Resveratrol and SIRT-1:

Chukwumah Y, Walker L, Vogler B, Verghese M. Changes in the phyto-chemical composition and profile of raw, boiled, and roasted peanuts. *J. Agric Food Chem.* 2007;55(22): 9266–9273.

Movahed A, Nabipour I, Lieben Louis X, et al. Antihyperglycemic effects of short term resveratrol supplementation in type 2 diabetic patients. *Evid Based Complement Alternat Med.* 2013;2013:851267.

Anisimova NY, Kiselevsky MV, Sosnov AV, Sadovnikov SV, Stankov IN, Gakh AA. Trans-, cis-, and dihydro-resveratrol: a comparative study. *Chem Cent J.* 2011; 5:88.

Chapter Six

Hormesis and Bioactives:

Mattson MP. Hormesis defined. Ageing Res Rev. 2008;7(1):1–7. doi:10.1016/j.arr.2007.08.007

Chirumbolo S. Hormesis, resveratrol and plant-derived polyphenols: some comments. *Hum Exp Toxicol.* 2011;30(12):2027-30.

Curcumin Bioactive:

Laxmidhar D, Manjula V. Long term effect of curcumin in restoration of tumour suppressor p53 and phase-II antioxidant enzymes via activation of Nrf2 signalling and modulation of inflammation in prevention of cancer. PLoS ONE. 2015;10(4): e0124000.

Curcumin and Inflammation:

Esatbeyoglu T, Ulbrich K, Rehberg C, Rohn S, Rimbach G. Thermal stability, antioxidant, and anti-inflammatory activity of curcumin and its degradation product 4-vinyl guaiacol. *Food Funct.* 2015; Mar;6(3):887-93.

Myrosinase:

Wang GC, Farnham M, Jeffery EH. Impact of thermal processing on sulforaphane yield from broccoli (Brassica oleracea L. var. italica). Journal of Agricultural and Food Chemistry. 2015;60:6743-6748.

Chapter Seven

Gene Expression and Methylation:

Moore LD, Le T, Fan G. DNA methylation and its basic function. *Neuropsychopharmacology.* 2012;38(1):23-38.

Glutathione

Pizzorno J. Glutathione! *Integr Med (Encinitas).* 2014;13(1):8-12.

Gibney ER, Nolan CM. Epigenetics and gene expression. *Heredity (Edinb).* 2010;Jul;105(1):4-13.

Methylation and cancer

Szyf M. The role of DNA hypermethylation and demethylation in cancer and cancer therapy. *Curr Oncol.* 2008;15(2):72-5.

Ehrlich M. DNA hypomethylation in cancer cells. *EPIGENOMICS VOL. 1, NO. 2 REVIEW.* Published Online:3 Dec 2009 https://doi.org/10.2217/epi.09.33.

Deeper Learning:

Bendall JK, Douglas G, McNeill E, Channon KM, Crabtree MJ. Tetrahydrobiopterin in cardiovascular health and disease. *Antioxid Redox Signal.* 2014;20(18):3040-77.

Eating Beets, Beet Greens for Better Methylation:

Obeid R. The metabolic burden of methyl donor deficiency with focus on the betaine homocysteine methyltransferase pathway. *Nutrients.* 2013;5(9):3481-95.

Chapter Eight

Omega 3s and Inflammation:

Calder P. Omega-3 fatty acids and inflammatory processes. *Nutrients.* 2010;2(3):355-74.

Patterson E, Wall R, Fitzgerald GF, Ross RP, Stanton C. Health implications of high dietary omega-6 polyunsaturated Fatty acids. *J Nutr Metab.* 2012;2012:539426.

Zárate R, El Jaber-Vazdekis N, Tejera N, Pérez JA, Rodríguez C. Significance of long chain polyunsaturated fatty acids in human health. *Clin Transl Med.* 2017; 6(1):25.

Vernekar M, Amarapurkar D. Diet-Gene Interplay: An Insight into the Association of Diet and FADS Gene Polymorphisms. *J Nutr Food Sci.*2016; 6:503.

De Boer AA, Monk JM, Liddle DM, Hutchinson AL, Power KA, Ma DW, Robinson LE. Fish-oil-derived n-3 polyunsaturated fatty acids reduce NLRP3 inflammasome activity and obesity-related inflammatory cross-talk between adipocytes and CD11b(+) macrophages. *J Nutr Biochem.* 2016;34:61-72.

Cardio protective effects of omega-3 fatty acids:

Layne J, Majkova Z, Smart EJ, Toborek M, Hennig B. Caveolae: a regulatory platform for nutritional modulation of inflammatory diseases. *J Nutr Biochem.* 2011;22(9):807-11.

Lee JY, Zhao L, Hwang DH. Modulation of pattern recognition receptor-mediated inflammation and risk of chronic diseases by dietary fatty acids. *Nutr Rev.* 2010;Jan;68(1):38-61.

Adkins Y, Kelley DS. Mechanisms underlying the cardioprotective effects of omega-3 polyunsaturated fatty acids. *J Nutr Biochem.* 2010;21(9):781-92.

General:

Kiecolt-Glaser JK, Glaser R, Christian LM. Omega-3 fatty acids and stress-induced immune dysregulation:

Olive oil:

Olive Oil Times Staff. "Keeping Olive Oil Fresh." *Olive Oil Times.* Mar. 19, 2010, www.oliveoiltimes.com/olive-oil-basics/keeping-olive-oil-fresh/878.

"What Effects the Quality of Olive Oil?" California Olive Ranch. www.californiaoliveranch.com/olive-oil-101/what-affects-the-quality-of-olive-oil/.

Chapter Ten

Federal government and omegas:

Food Labeling: Nutrient Content Claims; Alpha-Linolenic Acid, Eicosapentaenoic Acid, and Docosahexaenoic Acid Omega-3 Fatty Acids. A Rule by the Food and Drug Administration on 04/28/2014. https://www.federalregister.gov/documents/2014/04/28/2014-09492/food-labeling-nutrient-content-claims-alpha-linolenic-acid-eicosapentaenoic-acid-and-docosahexaenoic.

Guidance for Industry: Nutrient Content Claims; Alpha-Linolenic Acid, Eicosapentaenoic Acid, and Docosahexaenoic Acid Omega-3 Fatty Acids; Small Entity Compliance Guide. https://www.fda.gov/Food/GuidanceRegulation/GuidanceDocumentsRegulatoryInformation/ucm484250.htm .

Nutrient composition of foods:

Marles RJ. Mineral nutrient composition of vegetables, fruits and grains: The context of reports of apparent historical declines. *J Food Compos and Anal.* 2017; 56: 93-103.

What Americans eat vs. government recommendations:

Current Eating Patterns in the United States. https://health.gov/dietaryguidelines/2015/guidelines/chapter-2/current-eating-patterns-in-the-united-states/. Last Accessed Dec 2, 2018.

Chart: A Closer Look at Current Intakes and Recommended Shifts. https://health.gov/dietaryguidelines/2015/guidelines/chapter-2/a-closer-look-at-current-intakes-and-recommended-shifts/#table-2-1. Last accessed Dec 2, 2018.

Chart: State of the Plate 2015. Produce for Better Health Organization. http://www.pbhfoundation.org/pdfs/about/res/pbh_res/ State_of_the_Plate_2015_WEB_Bookmarked.pdf.

DHA and EPA's action in cells:

Sakai C, Ishida M, Ohba H, et al. Fish oil omega-3 polyunsaturated fatty acids attenuate oxidative stress-induced DNA damage in vascular endothelial cells. *PLoS One.* 2017;12(11).

Chapter Eleven

Health benefits of garbanzo beans:

Wallace TC, Murray R, Zelman KM. The Nutritional Value and Health Benefits of Chickpeas and Hummus. *Nutrients.* 2016;8(12):766.

Health benefits of lentils:

Ganesan K, Xu B. Polyphenol-Rich Lentils and Their Health Promoting Effects. *Int J Mol Sci.* 2017;18(11):2390.

Cynthia Chatterjee C, Gleddie S, Xiao CW. Bioactive Peptides and Their Functional Properties. *Nutrients* 2018;10(9):1211.

Health benefits of flaxseed:

Martinchik AN, Baturin AK, Zubtsov VV, Molofeev VIu. Nutritional value and functional properties of flaxseed. *Vopr Pitan.* 2012; 81(3): 4–10.

Kajla P, Sharma A, Sood DR. Flaxseed-a potential functional food source. *J Food Sci Technol.* 2014;52(4):1857-71.

Health benefits of sesame:

Namiki M. Nutraceutical functions of sesame: a review. *Crit Rev Food Sci Nutr.* 2007;47(7):651-73.

De Almeida A, Gouveia V, Cardoso C, et al. Effects of the intake of sesame seeds (*Sesamum indicum* L.) and derivatives on oxidative stress: A systematic review. *J Med Food.* 2016;19:4, 337-34.

Nagendra Prasad MN, Konsaur RS, Deepika SP. A Review on Nutritional and Nutraceutical Properties of Sesame. *J Nutr Food Sci.* 2012; 2(2):1-6.

Health benefits of sunflower seeds:

Faqir MA, Muhammad N, Muhammad IK, Shahzad H. Nutritional and therapeutic potential of sunflower seeds: a review. *British Food Journal.* 2012; 114(4):544-552.

Akrami A, Nikaein F, Babajafari S, Faghih S, Yarmohammadi H. Comparison of the effects of flaxseed oil and sunflower seed oil consumption on serum glucose, lipid profile, blood pressure, and lipid peroxidation in patients with metabolic syndrome. J Clin. *Lipidology*. 2018; 12(1):70–77.

Health benefits of seaweed and algae:

Wells ML, Potin P, Craigie JS, Raven JA, Merchant SS, Helliwell KE. Algae as nutritional and functional food sources: revisiting our understanding. *J Appl Phycol*. 2016;29(2):949-9.

Chapter Twelve

Digestive problems:

Konturek PC, Brzozowski T, Konturek SJ. Stress and the gut: Pathophysiology, clinical consequences, diagnostic approach and treatment options. *J Physiol Pharmacol*. 2011;62(6):591-9.

Esch T, Stefano GB, The neurobiology of stress management. *Neuro Endocrinol Lett*. 2010;31(1):19-39.

Stress and the gut:

Taché Y, Bonaz B. Corticotropin-releasing factor receptors and stress-related alterations of gut motor function. *J Clin Invest*. 2007;117(1):33-40.

Puterman E, Lin J, Krauss J, Blackburn EH, Epel ES. Determinants of telomere attrition over 1 year in healthy older women: stress and health behaviors matter. *Mol Psychiatry*. 2014;20(4):529-35.

Irritable Bowel Syndrome:

Sansone RA, Sansone LA. IRRITABLE BOWEL SYNDROME: Relationships with Abuse in Childhood. *Innov Clin Neurosci*. 2015;12(5-6):34-7.

Human Social Genomics:

Cole SW. Human social genomics. *PLoS Genet*. 2014;10(8):e1004601. Published 2014 Aug 28.

Slavich GM, Cole SW. The Emerging Field of Human Social Genomics. *Clin Psychol Sci*. 2013;1(3):331-348.

Michael Gershon Book and gut-brain connections:

Gershon, Michael. The second brain: A groundbreaking new understanding of nervous disorders of the stomach and intestine. Harper Perennial:

New York, NY: 1999. 336 pages.

Komaroff, Anthony L. "The gut-brain connection: Pay attention to your gut-brain connection – it may contribute to your anxiety and digestion problems." *Healthbeat,* Harvard Health Publishing, https://www.health.harvard.edu/diseases-and-conditions/the-gut-brain-connection.

"The brain-gut connection." *Healthy Aging,* Johns Hopkins Medicine. https://www.hopkinsmedicine.org/health/healthy_aging/healthy_body/the-brain-gut-connection. Last accessed Nov 18, 2018.

Butyrate and enterocytes:

Canani RB, Costanzo MD, Leone L, Pedata M, Meli R, Calignano A. Potential beneficial effects of butyrate in intestinal and extraintestinal diseases. *World J Gastroenterol.* 2011;17(12):1519-28.

Miron N, Cristea V. Enterocytes: active cells in tolerance to food and microbial antigens in the gut. *Clin Exp Immunol.* 2012;167(3):405-12.

The Gut and our brain:

Selhub EM, Logan AC, and Bested AC. Fermented foods, microbiota, and mental health: ancient practice meets nutritional psychiatry. *J Physiol Anthropol.* 2014;33(1): 2.

Vaginal delivery and immune health:

Neu J, Rushing J. Cesarean versus vaginal delivery: long-term infant outcomes and the hygiene hypothesis. *Clin Perinatol.* 2011;38(2):321-31.

Proton pump inhibitors:

Freedberg DE, Lebwohl B, Abrams JA. The impact of proton pump inhibitors on the human gastrointestinal microbiome. *Clin Lab Med.* 2014;34(4):771-85.

Reveles KR, Ryan CN, Chan L, Cosimi RA, Haynes WL. Proton pump inhibitor use associated with changes in gut microbiota composition. *Gut.* 2017;67(7):1369-1370.

Bacterial imbalance in the gut and weight control:

Turnbaugh PJ, Ley RE, Mahowald MA, Magrini V, Mardis ER, Gordon JI. An obesity-associated gut microbiome with increased capacity for energy harvest. *Nature.* 2006; 444: 1027-1031.

Kallus SJ, Brandt LJ.The intestinal microbiota and obesity. *J Clin Gastroenterol.* 2012;46(1):16-24.

Baothman OA, Zamzami MA, Taher I, Abubaker J, Abu-Farha M. The role of Gut Microbiota in the development of obesity and Diabetes. *Lipids Health Dis.* 2016; 15:108.

De Silva A, Bloom SR. Gut Hormones and Appetite Control: A Focus on PYY and GLP-1 as Therapeutic Targets in Obesity. *Gut Liver.* 2012;6(1):10-20.

Definition of probiotic:

Hill C, Guarner F, Reid G, Gibson GR, Merenstein DJ, Pot B,et al. Expert consensus document. The International Scientific Association for Probiotics and Prebiotics consensus statement on the scope and appropriate use of the term probiotic. *Nat Rev Gastroenterol Hepatol.* 2014;11: 506–514.

About fermented foods and probiotics:

Heller KJ. Probiotic bacteria in fermented foods: product characteristics and starter organisms. *Am J Clin Nutr.* 2001;73(2): 374s-379s.

Selhub EM, Logan AC, Bested AC. Fermented foods, microbiota, and mental health: ancient practice meets nutritional psychiatry. *J Physiol Anthropol.* 2014;33(1):2.

Chilton SN, Burton JP, Reid G. Inclusion of fermented foods in food guides around the world. *Nutrients.* 2015;7(1):390-404.

Shi Y, Zhao X, Zhao J, Zhang H, Zhai Q, Narbad A, et al. A mixture of Lactobacillus species isolated from traditional fermented foods promote recovery from antibiotic-induced intestinal disruption in mice. *J Appl Microbiol.* 2018;124(3):842-854.

Tamang JP, Shin DH, Jung SJ, Chae SW. Functional Properties of Microorganisms in Fermented Foods. *Front Microbiol.* 2016;7:578.

Marco ML, Heeney D, Binda S, Cifelli CJ, Cotter PD, Foligné B, et al. Health benefits of fermented foods: microbiota and beyond. *Curr Opin Biotechnol.* 2017; 44:94-102.

Natto, vitamin K, and bone health:

Adam J, Pepping J. Vitamin K in the treatment and prevention of osteoporosis and arterial calcification. *Am J Health Syst Pharm.* 2005;62(15):1574-1581.

Arunakul M, Niempoog S, Arunakul P, Bunyaratavej N. Level of under-carboxylated osteocalcin in hip fracture Thai female patients. *J Med Assoc Thai.* 2009; 92(Suppl5):S7-11.

Yaegashi Y, Onoda T, Tanno K, Kuribayashi T, Sakata K, Orimo H.

Association of hip fracture incidence and intake of calcium, magnesium, vitamin D, and vitamin K. *Eur J Epidemiol.* 2008;23(3):219–225.

Kaneki M, Hodges SJ, Hosoi T, Fujiwara S, Lyons A, Crean SJ, et al. Japanese fermented soybean food as the major determinant of the large geographic difference in circulating levels of vitamin K2: possible implications for hip-fracture risk. *Nutrition.* 2001;17(4):315-21.

Ikeda Y, Iki M, Morita A, Kajita E, Kagamimori S, Kagawa Y, Yoneshima H. Intake of fermented soybeans, natto, Is associated with reduced bone loss in postmenopausal women: Japanese population-based osteoporosis (JPOS) Study. *J. Nutr.* 2006;136(5): 1323-1328.

Fermented foods benefits:

Marco ML, Heeney D, Binda S, Cifelli CJ3 Cotter PD4 Foligné B ,et al Health benefits of fermented foods: microbiota and beyond. *Curr Opin Biotechnol.* 2017; Apr;44:94-102.

Heart health and fermented foods:

Gast GC, de Roos NM, Sluijs I, Bots ML, Beulens JW, Geleijnse JM,A high menaquinone intake reduces the incidence of coronary heart disease. *Nutr Metab Cardiovasc Dis.* 2009; 19(7):504-10.

Geleijnse JM, Vermeer C, Grobbee DE, Schurgers LJ, Knapen MH, van der Meer IM, et al. Dietary intake of menaquinone is associated with a reduced risk of coronary heart disease: the Rotterdam Study. *J Nutr.*2004;134(11):3100-5.

Walther B, Karl JP, Booth SL, Boyaval P. Menaquinones, bacteria, and the food supply: the relevance of dairy and fermented food products to vitamin K requirements. *Adv Nutr.* 2013;4(4):463-73.

Knapen MH, Braam LA, Drummen NE, Bekers O, Hoeks AP, Vermeer C.Menaquinone-7 supplementation improves arterial stiffness in healthy postmenopausal women. A double-blind randomised clinical trial. *Thromb Haemost.*2015;113(5):1135-44.

Dr. Miller's book:

Miller, Daphne. *The Jungle Effect: Healthiest diets from around the world.*William Morrow Paperbacks: New York, NY: 2009.384 pp.

More benefits of fermented foods:

Bondia-Pons I, Nordlund E, Mattila I, et al. Postprandial differences in the plasma metabolome of healthy Finnish subjects after intake of a sourdough fermented endosperm rye bread versus white wheat bread. *Nutr J.* 2011;10:116.

Ejtahed HS, Mohtadi-Nia J, Homayouni-Rad A, Niafar M, Asghari-Jafarabadi M, Mofid V. Probiotic yogurt improves antioxidant status in type 2 diabetic patients. *Nutrition.* 2012;28(5):539-43.

Selhub EM, Logan AC, Bested AC. Fermented foods, microbiota, and mental health: ancient practice meets nutritional psychiatry. *J Physiol Anthropol.* 2014;33(1):2.

Bose S, Kim H. Evaluation of in vitro anti-Inflammatory activities and protective effect of fermented preparations of rhizoma atractylodis macro-cephalae on intestinal barrier function against lipopolysaccharide insult. *Evid Based Complement Alternat Med.* 2013; 363076.

Parkar SG1, Trower TM, Stevenson DE. Bioactives and Gut Health:Fecal microbial metabolism of polyphenols and its effects on human gut micro-biota. *Anaerobe.* 2012; 23:12-19.

Chapter Thirteen

Resources for learning about fermentation, techniques and recipes:

Cultures for Health. www.culturesforhealth.com.

Body Ecology. www.bodyecology.com.

Nourished Kitchen. www.nourishedkitchen.com.

Mcgruther, Jenny. *The Nourished Kitchen.* Ten Speed Press/Penguin Random House: New York, NY: 2014, 320 pp.

Rawlings, Deirdre. *Fermented Foods for Health.* Fairwinds Press:Vancouver, BC: 2013, 208pp.

Shockey, Kirsten and Shockey, Christopher. *Fermented Vegetables: Creative Recipes for Fermenting 64 Vegetables & Herbs in Krauts, Kimchis, Brined Pickles, chutneys, Relishes & Pastes.* Storey Publishing: North Adams, MA: 2014, 376pp.

Appendix A: Genomic Testing. For Me or Not for Me?

p.1: N of 1 trial:

Lillie EO, Patay B, Diamant J, et al. The n-of-1 clinical trial: the ultimate strategy for individualizing medicine? Per Med. 2011; 8(2):161-173.

Appendix B: SIRT-1 and PPARG

SIRT-1:

https://www.ncbi.nlm.nih.gov/pmc/articles/PMC2853213/

https://www.ncbi.nlm.nih.gov/pmc/articles/PMC3545644/

PPARG:

https://www.ncbi.nlm.nih.gov/pubmed/20216208

https://academic.oup.com/abbs/article/45/1/51/1158/
SIRT1-and-energy-metabolism

About the Author

Amanda Archibald, RD, developed the concept of culinary genomics over a career spent educating nutritionists, chefs, and food service and food industry professionals around the world. Her hands-on workshops have helped thousands make the real-world connection between genes, what and how we eat, and the optimal functioning of our body systems. She is the founder of The Genomic Kitchen, a pioneering consultancy, online education program, and nutrition counseling service that makes optimal health accessible through its merging of genomic and nutrition science with the culinary arts. The UK native turned U.S. citizen is an avid cyclist and a passionate participant in all snow sports and currently resides in Wisconsin